U0880742

Urban Square Landscape

最新城市广场景观

何心 译

高迪国际出版有限公司 编
大连理工大学出版社

图书在版编目 (CIP) 数据

最新城市广场景观 : 英汉对照 / 高迪国际出版有限公司编 ; 何心译. —大连 : 大连理工大学出版社 , 2013.2
ISBN 978-7-5611-7468-5

Ⅰ. ①最… Ⅱ. ①高… ②何… Ⅲ. ①广场—景观设计—世界—图集 Ⅳ. ① TU984.18-64

中国版本图书馆 CIP 数据核字 (2012) 第 289115 号

出版发行：大连理工大学出版社
（地址：大连市软件园路 80 号 邮编：116023）
印　　刷：利丰雅高印刷（深圳）有限公司
幅面尺寸：245mm × 320mm
印　　张：21
插　　页：4
出版时间：2013 年 2 月第 1 版
印刷时间：2013 年 2 月第 1 次印刷
策划编辑：袁　斌　刘　蓉
责任编辑：刘　蓉
责任校对：王丹丹
封面设计：高　迪

ISBN 978-7-5611-7468-5
定　　价：320.00 元

电话：0411-84708842
传真：0411-84701466
邮购：0411-84703636
E-mail:designbooks_dutp@yahoo.cn
URL:http://www.dutp.cn

如有质量问题请联系出版中心：（0411）84709246　84709043

PREFACE 1

序言一

Zoltán Stéhli
landscape architect
GARTEN STUDIO LTD.

Today landscape architecture tries to transform cosmopolitan inputs into local outputs. Because of the wider range of effects through the internet, the design become more complex. The world of architecture, design, fashion, media and arts are interweaving in the established landscapes.

It is always interesting to find the local patterns of various spaces which create the connections to its surroundings, and use new possibilities e.g. computer graphics so our landscape is a bit more like a body than just surface.

If our extracting process is successful, the result will be quality public squares which become important scene of social life, and in good cases identify good spots in cityscape.

Good tendence of renewal and re-functioning downtown open spaces, gives more possibility for landscape architects to respond to public demands. The human scale usability, the green in the name of the sustainable development became as, or more important organizing principals of urban planning, than traffic or utility. Most of the projects put the green idea forward and the visible part of this is the exterior architecture.

Our current responsibility is to use the positive turn, to show the economic, social and environmental advantages of landscape design. The designed public squares have to stand the test of time, both in design and technical way. In the future there will be various frame systems for quantifying the useful effects, categorizing the investments, while the researches in this multidisciplinary topic will influence the design process too.

Next to economic and environmental benefits, new aspects of the social part are in our focus. A kind of a "vitality factor" measures the success of new public areas. In most city centers, pedestrian streets or mixed-use squares take place from vehicle traffic space. People start to use valuable city spaces for living instead of transportation. Public squares become part of their identity and an important scene of their social relations.

当今，景观设计的趋势是将全球性的流行元素转化为具有本地特色的设计。通过互联网的巨大影响，设计规划变得十分复杂。世界范围内的各种元素，包括建筑、设计、时尚、媒体甚至是艺术范畴，都将混合交织在已经完成的景观设计中。

尽力发掘本地不同空间的规划模式是一种十分有趣的方式，这种规划模式利用包括计算机制图在内的各种新的方法和手段，将空间和周边的环境相互连接，因此，我们所设计的景观不仅仅局限于简单的表面，而更像是一个有机的整体。

如果我们设计规划的过程十分顺利成功，那么结果就是打造出一个质量上乘的公共广场，为周边的社会生活提供一个十分重要的活动场地，一个可以领略美好的城市风光的所在。

对于城市公共空间的重建以及功能重置来说，景观设计师们有更多的机会重新赋予公共空间新的功能，从而满足公众的需求。更多人性化的适用设计以及可持续发展的绿色空间等，在城市规划的组织构架中被赋予了更高的设计地位，远高于公共交通或者其他公共事业规划。大多数项目极为推崇绿色理念，整个建筑的外部形貌都淋漓尽致地体现了这种规划意识。

当前，我们对于设计规划的主要责任是尽量利用各种积极的因素，在设计景观作品中体现经济、社会以及环境的有利地位。我们所设计的公共广场，在设计和技术方面都要经得起时间的考验。未来将会出现不同的框架设计系统，针对有用的设计动产进行定量分析，针对设计投资进行分门别类，同时，不同学科、不同部门的研究人员也将对设计过程产生更多的影响。

除了经济与环境方面的利益，我们还十分关注景观设计的社会效益。我们提出了一个“活力因子”的概念，用以评价一个新的公共空间设计成功与否。对于大多数城市中心来说，步行街区或者多用途广场通常由车辆交通的空间改建而成。人们开始意识到，应该使用更多有价值的城市空间为居民生活服务，而不是仅仅侧重于交通。公共广场开始印上城市居民的烙印，逐渐成为城市居民社会交往的重要场所。

PREFACE 2

序言二

Hanneke Kijne

co-owner & landscape architect

HOSPER landscape architecture and urban design

如何合理利用公共空间，尤其是那些位于繁华都市中现有的公共空间，已经成为目前人们持续关注的话题。需要通过详细谨慎的规划设计来完成改建，从而满足这种强烈的现实需求。我们工作室大多数的作品主要集中于花园、广场、公园或者整个公共空间网络的设计和布局。当然，这种开放式空间的设计风格始终贯穿于我们的设计作品中，包括居民区、工厂、景观或者整个地区的规划设计项目等等。

我们设计规划的主要目的是清晰地区分空间，突出主要的空间功能的概念，辅以各种细节上的精心制作。空间的功能性对于设计来说十分重要，而一个项目的设计规划绝不仅仅是将各种需求转化为功能空间的简单加和。有些时候，某些不寻常甚至是夸张的设计元素，会赋予空间一种健康的张力，一种如画的诗意。

每个问题及项目的构想和规划都会根据具体的条件和环境不断地变化，同样，不断变化的还包括项目的社会背景以及空间的社会地位等等。因此，每个项目都始于对各种问题的分析，乃至于对问题背后的问题的分析。我们力图深入细致地掌握项目规划的要点，因为只有做到敏锐而详细的规划才能获得必要的答案。

我们接受的项目作品范围广泛，风格不尽相同。我们经常与其他部门的同事合作，包括技术人员、生态学者、计划经济执业者、建筑师和艺术家等等。而我们在公司中以团队的方式工作的时候，总会连续不断地对项目任务进行讨论。这就需要给予每个团队成员相应的授权。

The use of space has become increasingly intensive and space which has been built predominates in our urban areas. This very fact results in a demand for strongly and carefully designed spaces upon which no building has taken place. Much of our work is directed towards the layout of gardens, squares, parks or of whole networks of public space. But, also when we are involved in making plans for new residential and work areas, landscapes or regions, it is the open space which inspires the foundation of our designs.

Our ambition is to produce lucid spaces in which the major forces are recognizable and all elaboration of detail forms an essential part of the concept. Functionality is very important but a design in its totality is more than the sum of a programme of demands translated into space. Elements which are unusual or even cocksure can grant a space a healthy tension and poetic significance.

Not only do the formulation of the problem and the program change for every assignment but so do the social context and the spatial situation. This is the reason why the design process begins with an analysis of the question, or even better, the question lying behind the question. We attempt to penetrate to the gist of the formulation of the problem, because it is only to a keenly formulated question that a sufficient answer can be given.

The commissions we receive are often wide-ranging and they differ widely in character. We have become used to working with personnel from other disciplines, such as technicians, ecologists, practitioners of planned economy, architects and artists. Just as when we work within our firm, working within a planning team requires a continuous and on-going debate about the assignment. This makes demands on the authority granting the commission as well as on all the other members of the planning team.

PREFACE 3

序言三

Professor Tone Lindheim, ILP, University of Life Sciences

Bjørbekk & Lindheim Landscape Architects

Why is it so important for a town to have good urban spaces? Comfortable and beautiful public places in towns can give just the breathing space that is needed in a densely packed town. The urban space can give an opening in the built environment, and make it possible to see a little more of the sky. Such openings in the urban fabric give valuable views from buildings nearby, and not least give people who spend time in the town a small breathing space, a valued experience in contrast to the tiresome walking up and down urban streets.

An urban space gives an opportunity to halt for a moment, to sit down, to breathe calmly abd to take a break from a hectic urban schedule. Urban spaces provide places to gather, to see and to be seen. They can be used as political arenas, as places for the expression of opinions, for manifestations and demonstrations. And they can be used for social and cultural activities, for performances, for film shows, for organized as well as spontaneous events and expressions.

Urban spaces have also, through history, been important places for the purchase and sale of goods and services. Farmers have come to town with their products, fishermen have docked along the quays, and craftsmen have offered their products in the market.

Urban spaces often provide with valuable green elements to the urban context, much appreciated by urban users. Trees and bushes may make green oases in an otherwise highly trafficked, noisy and dusty town. Lawns may invite people to settle down, stretch out, romp around, play and frolic, remove shoes, and take a small break from daily chores.

Flowers can make beautiful items in urban spaces. Such organic elements help to bind dust, give out humidity and make a favourable environment for insects, birds and other animals. In an urban context, this may add substantially to the experience – hearing the chirps of birds, the humming of insects, sensing the changing of the seasons through the leafing of trees and bushes, the early flowering of tulip bulbs, the growth of large trees giving shade to grassy patches and walkways. And later, autumn comes with splendid colours, and then winter with interesting contours and curious silhouettes.

Such green elements have an importance far beyond the pragmatic uses and aesthetic value. A green environment contributes positively to an ecological context. It gives permeable surfaces so that water finds a slower and more natural way down into the ground. This can give better growth conditions for biodiversity within a densely built-up urban environment.

Good urban spaces make an invitation to social uses. It should be possible to sit down, take a small rest, have a good time alone or with others. Whether the seating should be located in the sun or the shade depends on the local climatic conditions. In Scandinavia it is often cold and dark for long periods of the year. Here it is important that people may sit and enjoy the sun whenever he is out. In other parts of the world, deep shade for cooling down may be more attractive. The use of water may give an attractive freshness in an urban space. It may be pleasant to experience the sounds of rippling, flowing, gushing, dripping, gurgling water. Or water may be completely still, reflects the sky and the large trees overhead.

If urban spaces are to function well at all times of day and night, they need good illumination. The light source should not blind people, but give security, focuses on important elements and makes it possible to use the spaces after dark.

A good urban space invites different uses throughout the day, week and year. An exemplary space appeals to different types of users – children, youth, elderly, and people of different social background. The spending of public funds on urban spaces can be more easily justified if the use is diverse, inclusive and intense.

When designing a public space, it is important to listen to its history, to sense the identity of the place, to understand the needs and desires of the users and to be aware of the social and physical context. High quality in design, detailing, choice of materials and construction is a characteristic of a good urban space. Such spaces should be solid, able to stand a punch and they should age with dignity.

对一个城市来说，如何规划设计出功能健全的都市公共空间，是一个十分重要的课题。舒适美丽的城市公共空间，是人潮拥挤的都市中所必需的休闲场所，是钢筋混凝土建筑环境所围绕的开放空间，在这里能够欣赏到都市中难得的广阔天空。广场为都市中的公众提供休闲场所之余，还为周边建筑中的人们提供赏心悦目的开放式景观，这是一种完全不同于典型的城市街道中忙忙碌碌、人来人往的体验和感受。

都市公共空间为公众提供了一个休闲小憩的机会和空间，或安然静坐，或平静呼吸，从繁忙的都市生活中解脱出来，享受短暂的悠闲时光。公共空间将周边空间整合在一起，功能灵活，可用作政治演说场地，也可作为表达观点的场所，同时也可以为社会文化活动、公共演出、电影放映以及自发的事件或组织活动提供场地服务。

从历史的角度来看，都市公共空间一直是商业、零售业以及服务业的重要场所。农夫、渔夫、手工业者们都会带着产品或者收获来市场出售。

城市空间通常为都市公众以及周边地带提供了大量宝贵的绿色元素。在交通拥堵、到处充斥着噪音和灰尘的都市中，树木和灌木带尤其珍贵，如同沙漠中的绿洲。即使是小小的一片绿色的草坪，人们也可以在上面静坐，伸展四肢，嬉笑玩闹，游戏狂欢，甚或除去鞋袜，在日常的喧嚣中享受难得的片刻悠闲。

鲜花能为都市空间带来美丽，这些植物元素能够吸纳灰尘，提高空气湿度，为昆虫、鸟类以及其他动物提供一个适宜的生存环境。对于都市生活来说，这无疑是一种难得的体验——聆听鸟儿的鸣叫和昆虫的哼唱，从片片树叶中感受季节的变迁，欣赏郁金香初放的柔美，享受默默生长的大树下阴凉的草坪和人行步道。随后，是秋天带来的令人目眩的五彩缤纷和浪漫多情，是冬天带来的让人兴致盎然的萧索宁静。

这些绿色元素所包含的意义远远超出所谓的实际用途和景观价值。绿色的环境对于一个生态系统来说意义非凡。例如，降水可以通过利于渗透的地表，以更自然的方式慢慢地渗透到更深层的土壤中，能够为人口密集的都市环境中的生物多样性提供更好的生长条件。

良好的都市开放空间还兼具一定的社交功能，人们在这里可以独自静坐，也可三五成群，享受短暂的娱乐休闲时光。可以根据当地的气候条件，决定座椅的安放位置，或是阳光下，或是树荫里，不一而足。在斯堪的纳维亚半岛，一年中的很长时间天气寒冷阴暗，人们能够坐在户外享受阳光，无论何时都是一件十分惬意并且重要的事情。而在世界的其他地区，人们可能更愿意坐在树荫下享受阴凉。在城市空间中合理地规划利用水系景观，是另一件令人赏心悦目的事情。人们可以静静地聆听不同水系带来的天籁之音，或小流潺潺，或迸流喷涌，或水滴叮咚，或清流汩汩，又或一池碧水寂静无声，倒映着蓝天、白云、青草、绿树。

当然，如果都市的公共空间入夜开放，则需要良好的照明系统。在不妨碍公众视线的前提下，灯光必须足够明亮，通常直射在一些重要的建筑物或景观上，在入夜后为公众提供一个十分具有安全感的活动空间。

规划设计合理的都市空间，成年累月、每时每刻都在提供服务，满足公众不同的需求。典型的公共空间能够吸引各式各样的群体，包括孩子、年轻人、老人以及来自社会各界具有各种不同社会背景的人们。如果都市公共空间功能多样，用途广泛，使用频率高，那么公众当然可以认为用于这里的公共资金十分合理。

在公共空间的规划设计过程中，最为关键和重要的问题包括追溯都市的历史，感受当地的风俗，理解公众的需求和愿望以及相关的社会背景。高质量的设计方案，细致入微的操作细节以及精心选择的建筑材料，是造就一个优秀的城市空间的必要元素。这样的优秀空间应该是坚实耐用，经得起历史的冲击，始终保持独有的尊严，历久弥新的。

CONTENTS

目 录

BGU UNIVERSITY ENTRANCE SQUARE & ART GALLERY

BGU 大学入口广场和美术馆

Beersheba, Israel

The Deichmann square and the Negev Gallery constitute a link between Ben-Gurion University campus and the city of Beersheba.

The square is located at the west gate of the campus, surrounded by some buildings and the presently constructing Negev Gallery. The square offers an outdoor space of cultural and social activities for students and citizens.

The square is bordered by the elongated gallery, facing both the city and the campus. Straightly against the city, the gallery's continuous facade (160 meters in length) unifies the heterogeneous appearance of the various buildings behind the gallery into a cohesive urban unit. The sculpture garden defines the boundary of the city, creating a green edge to the campus. The two-story high monolithic body of exposed concrete emerges from lawny topography of the northern part of the campus and hovers above an entrance courtyard in the southern part, where it appears to be leaping towards the urban space.

The gallery hosts exhibition spaces, museology faculty, workshops and auditorium contributing to the outdoor activities on Deichmann square. Since the square is designed to accommodate large parties for youth and students, the foremost scheme is to allocate limited areas for vegetation. The design of the square connects various elements of the surrounding buildings both physically and visually, accentuating their common features.

The square appears as a carpet of integrated strips of concrete paving, vegetation and lighting with concrete benches and trees scattered randomly. The strips of vegetation consist of lawn, Equisetopsida and seasonal plants.

site plan

• **Area** - 4,500 m² • **Photographer** - Sharon yeari

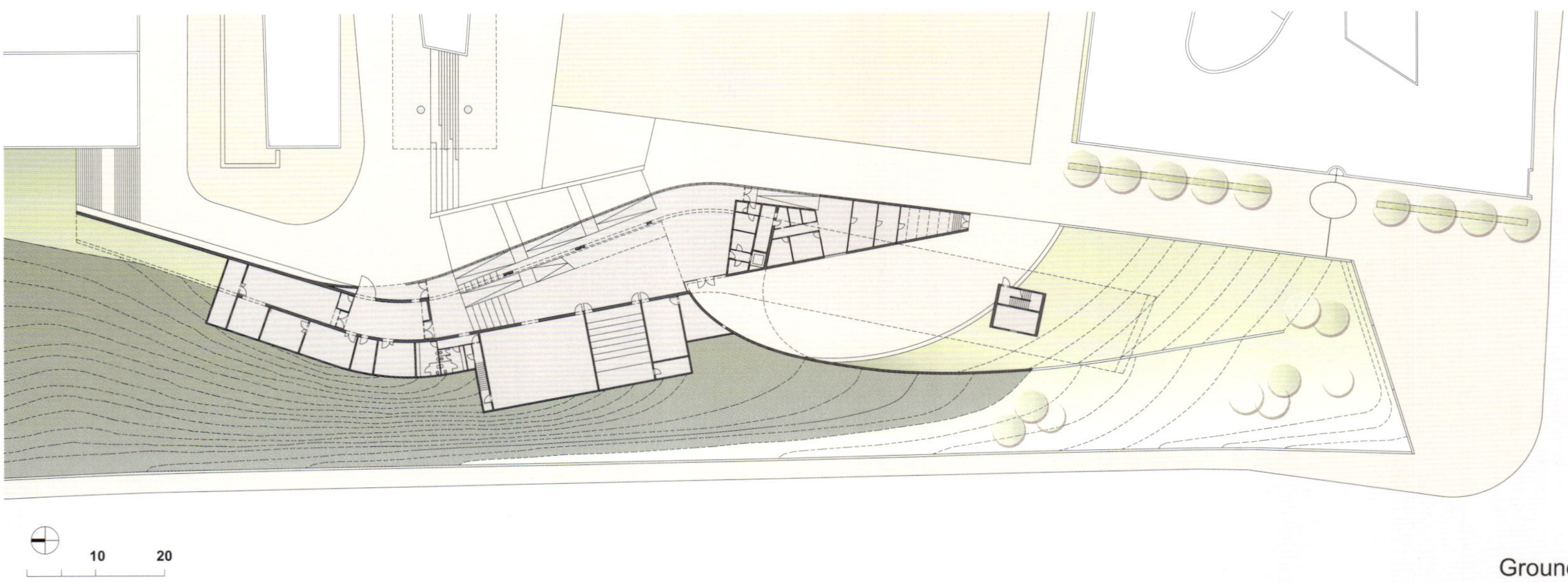

Ground Level

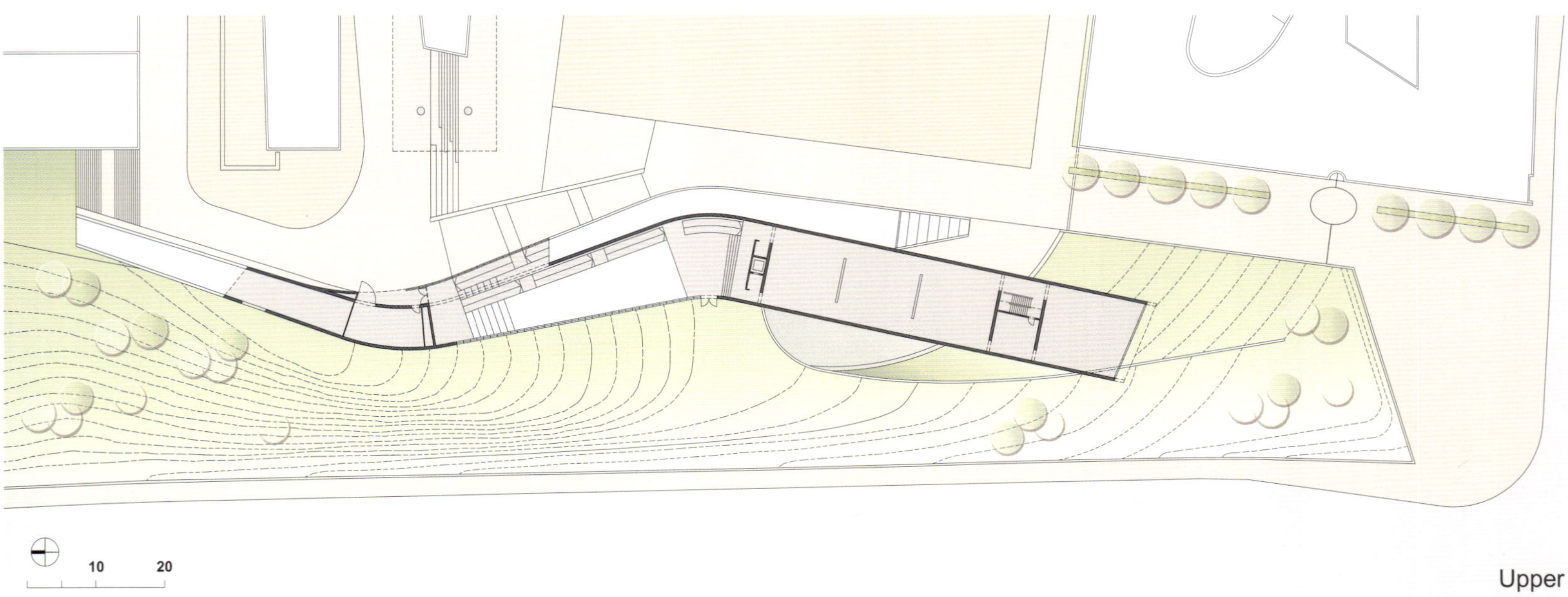

Upper Level

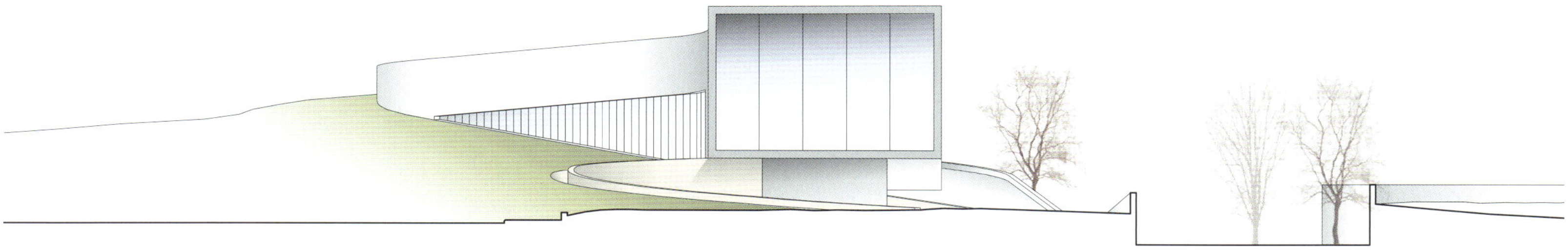
Facade

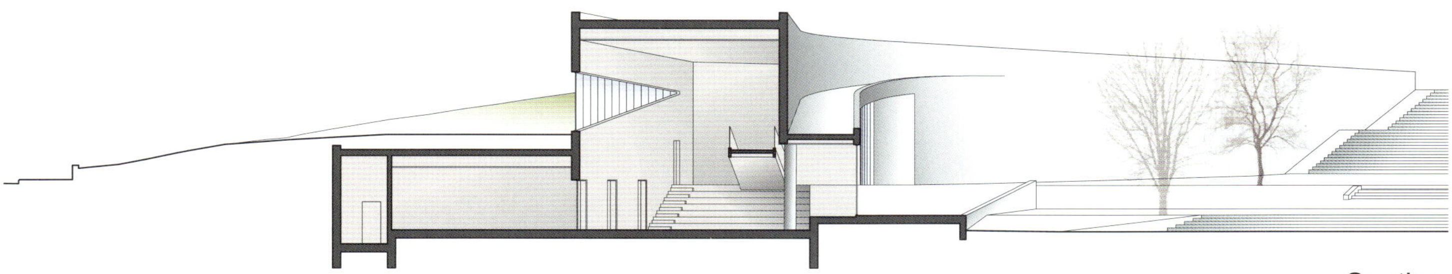
Section

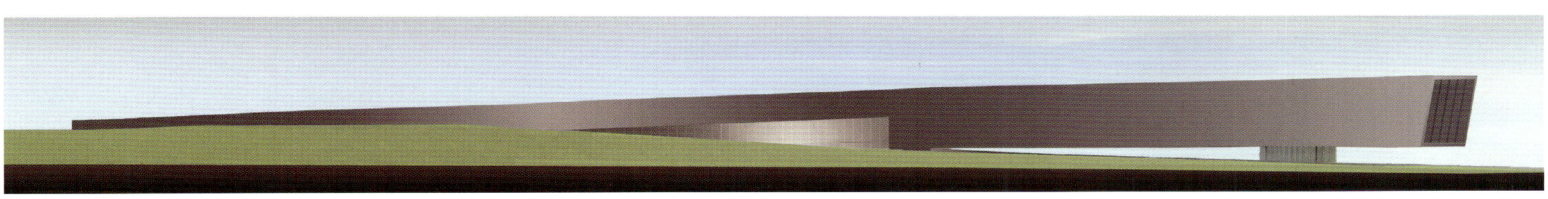
City Facade

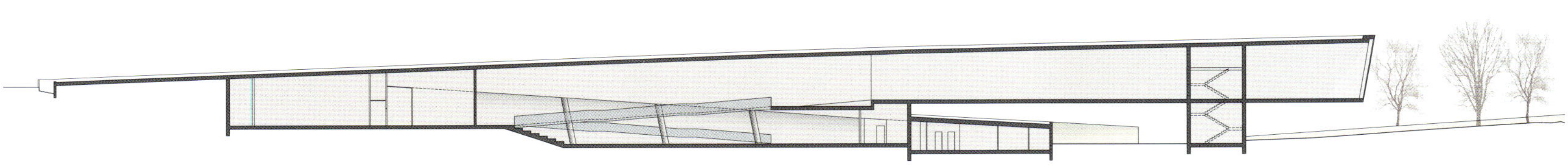
Long Section

Campus Facade

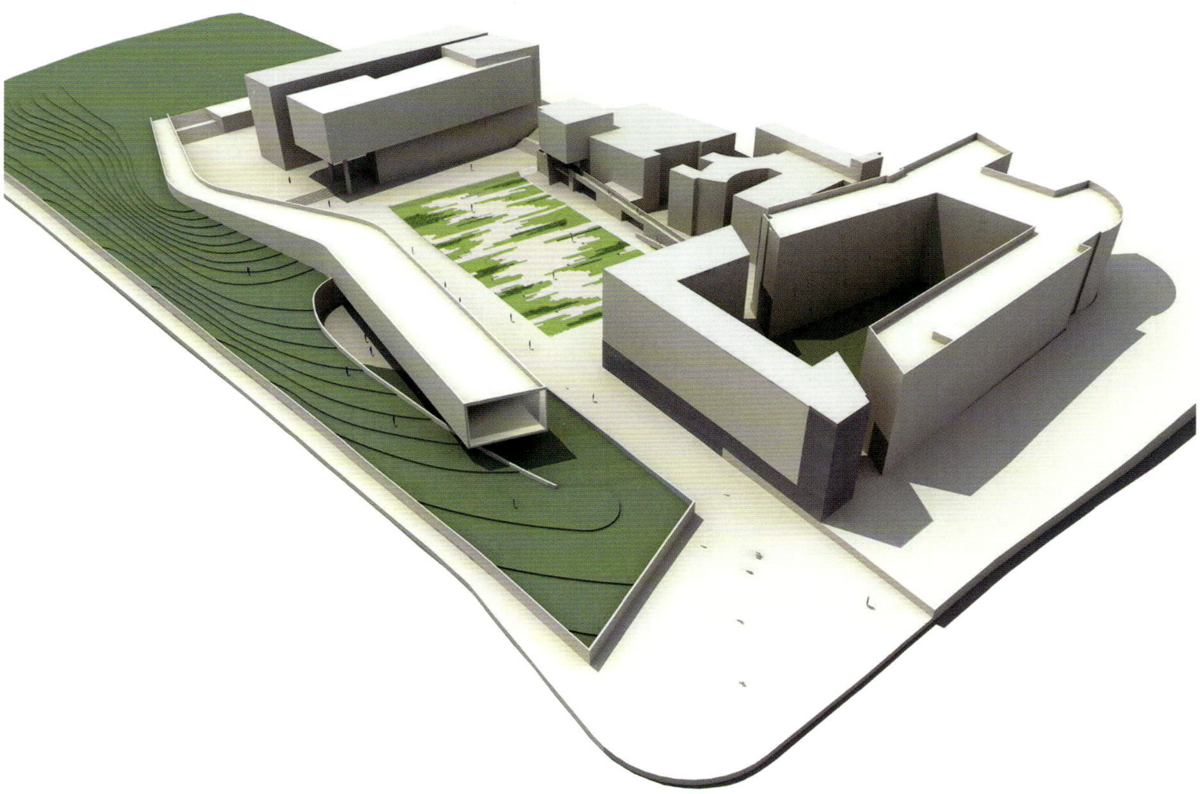

戴希曼广场和内盖夫美术馆是连接本·古里安大学校园和贝尔谢巴市之间的纽带。

广场位于校园的西门，被一些建筑物和在建的内盖夫美术馆包围。它为学生和市民提供了户外的社会文化活动空间。

广场与狭长结构的美术馆相邻，面朝城市和校园。正对着城市的是美术馆绵延的外墙（长 160 米），与美术馆后身形态各异的建筑浑然一体，构成一个紧凑的城市空间。雕塑公园界定了城市的边界，并为校园搭建了一条绿色的边缘线。两层高的露石混凝土结构建筑矗立在校园北部的草坪上，仿佛盘旋在南部庭院入口的上方，俨然一副要纵身一跃跳到市区的架势。

美术馆集展览、博物馆教学、讲习班和礼堂功能为一体，促进了戴希曼广场户外活动的开展。由于广场的设计旨在为年轻人和学生提供聚会的场所，所以，首选方案就是覆盖有限的植被。广场的设计在外观和视觉上结合了周围现存的混凝土建筑的诸多元素，同时也强调了它们的共性。

广场犹如混凝土路面上的一张绿色地毯，植物、灯光、石凳和树木都以长条状分布于广场之上。植被由草坪、木贼纲植物以及其他季节性植物组成。

FONTANA SQUARE IN QUINTO DE STAMPI

Rozzano, Milan, Italy

昆托斯塔皮的丰塔纳广场

Urban design for a public square in Rozzano, a suburb of Milan, offers the local community a flexible and shared outdoor space.

The aim of this project was to create a new, flexible and welcoming landscape for the local neighborhood which would satisfy the community's complex and continuously changing needs. Labics' intention was to make a space which would also trigger new, unplanned uses whilst retaining a strong local identity. A public consultation helped the municipality to define a very precise brief, accommodating the many different requirements and aspirations of the local community.

The geometrical pattern of the landscape is based on a dense grid of "golden rectangles" (i.e. rectangles with side lengths in the golden ratio of 1:1.6)and the dimensions of these determine every element of the square from the planting to the paving. A system of triangular shapes inside this orthogonal pattern helps to define the various natural and artificial surface treatments within the landscape, which include water, stone, lawn, shrubs and planting beds. A variety of trees, including cherries, pears, birches and acacias, have been planted to ensure a changing display of blossom and colour throughout the year. The paving is composed of a variety of materials ranging from local stone to iroko timber and concrete. To emphasize the lightly undulating topography of the square, the landscape has been artificially "bended" to create a dynamic space.

A contemporary sculpture which acts as a gateway to the neighborhood, benches and a pavilion/info point populate the square, providing spaces for play, relaxation and interaction.

· **Architectural Team** - Giorgio Pasqualini · **Model** - Susan Berardo, Gaia Maria Lombardo · **Area** - 6,200 m^2 · **Photographer** - Luigi Filetici

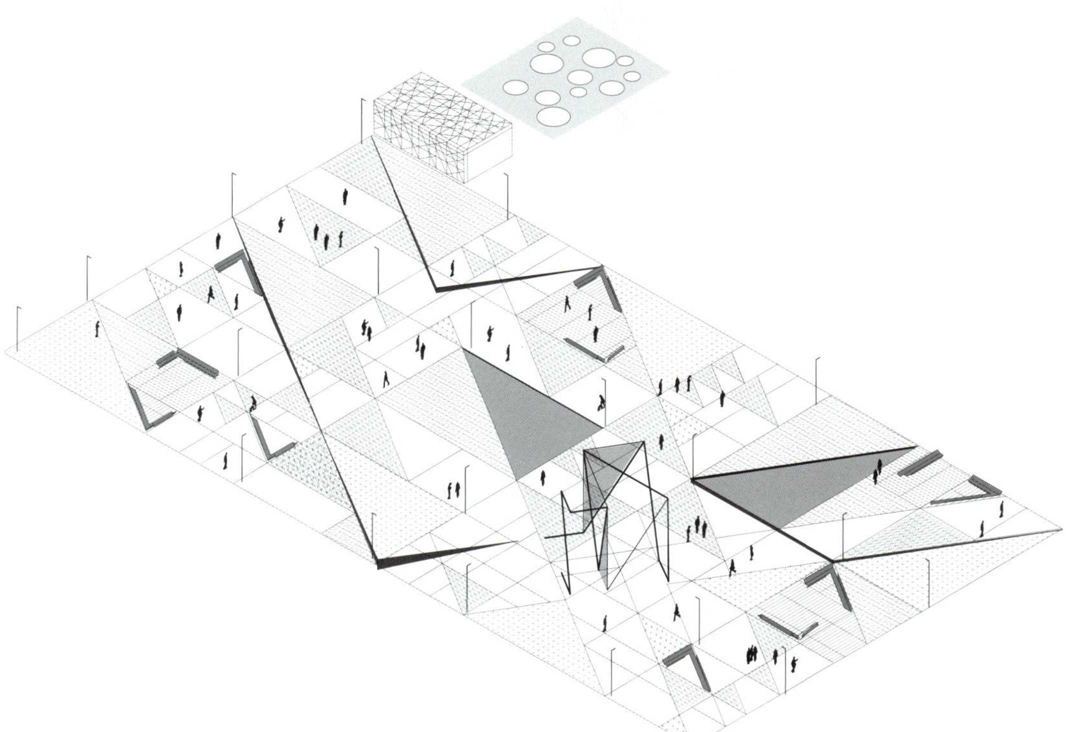

丰塔纳公共广场位于米兰郊区的罗扎诺地区，为当地社区提供了一个市民共享的室外空间。

广场项目的主要目的是为满足当地社区不断发展的综合需要，为当地居民建造一个崭新灵活的公共空间，一个充满好客气息的场景。拉比克斯建筑事务所负责整个广场的设计规划，设计目的是打造一个崭新的并能够根据需要灵活变化的空间，同时镌刻上深深的当地风格的烙印。为了满足当地居民的不同需求和愿望，当地政府组织了一场公开调查，经过磋商，最终确定了一个精确而简洁的方案。

广场的几何外形以黄金矩形的格子为基础，即矩形的长宽比遵循黄金比例 1:1.6，无论是植物的种植方式还是石块小路的铺设，这种外形的应用贯穿整个广场的每一个细节。矩形内部是一系列三角造型，用以区分各式不同的自然或人工表面处理方式，包括水系、石块造型、草地、灌木以及种植床等等。广场上种植了多种树木，有樱桃树、梨树、梓树、阿拉伯胶树等等，这种多样化的树种选择可以保证广场上一年四季绿意盎然，生机勃勃。人行小路由当地石材、绿柄桑木以及混凝土混合铺设而成。为了突出当地略带丘陵式的地理风貌，设计师们还特别选用了人工弯曲的手法，在广场上营造出一个颇具动感的空间。

广场入口处矗立着一座现代化风格的雕塑，长椅和亭廊遍布整个广场，为居民和游客提供了足够的休闲、娱乐和玩耍的互动空间。

11.34
10.05
10.94
6.53
38.86

Banca Intesa

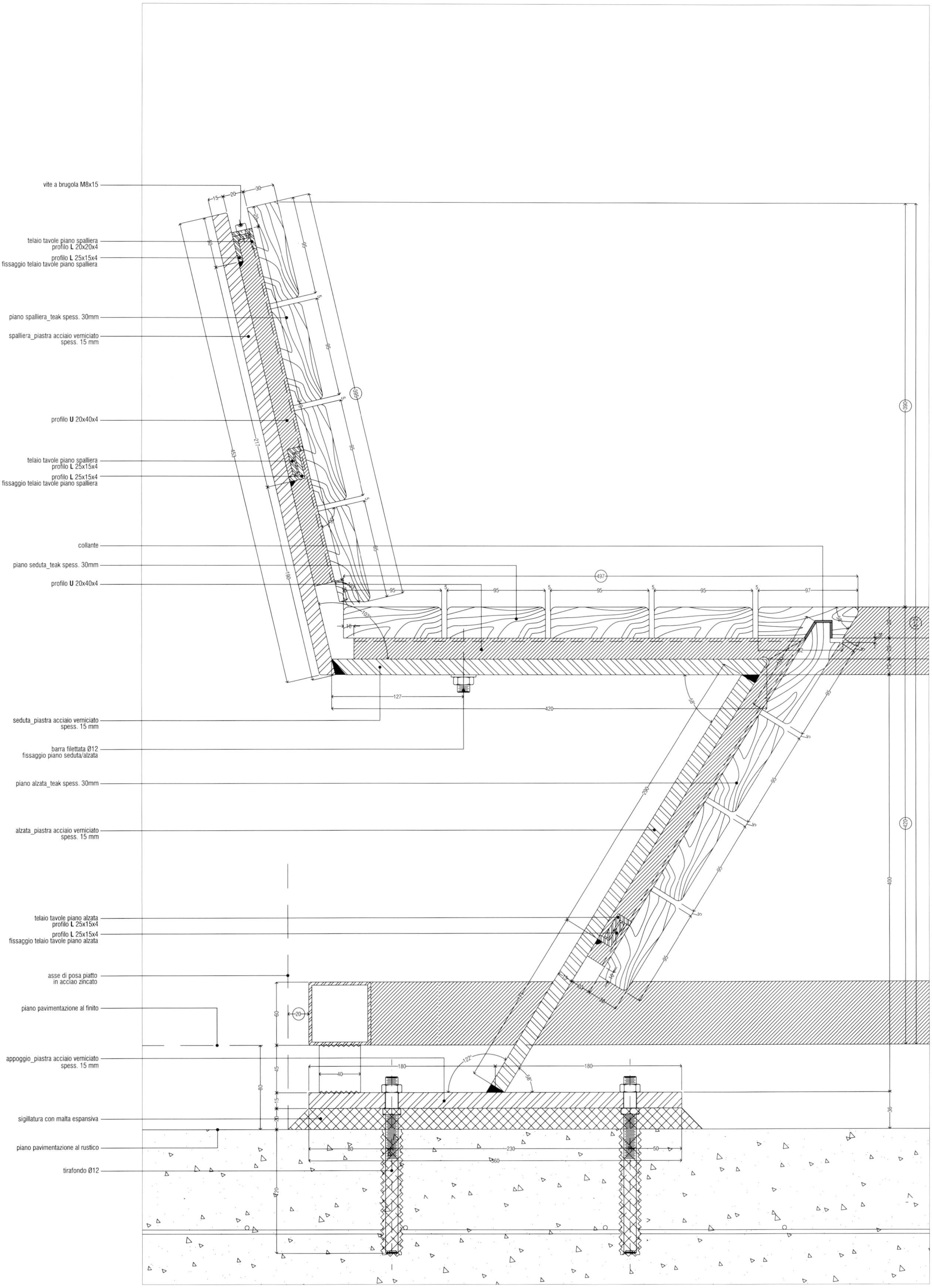

SEZIONE PANCA tipo A_ scala 1:2

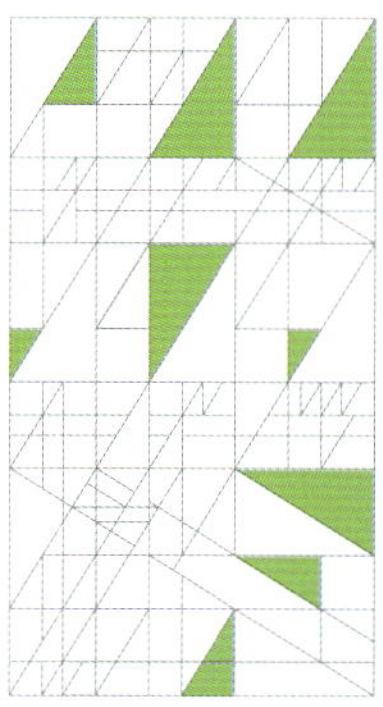

Prato

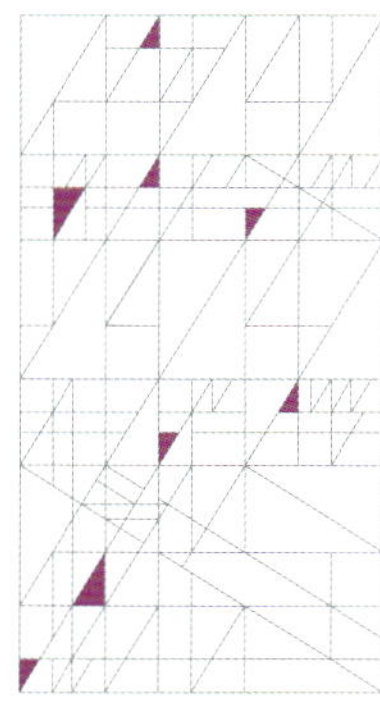

Fiori

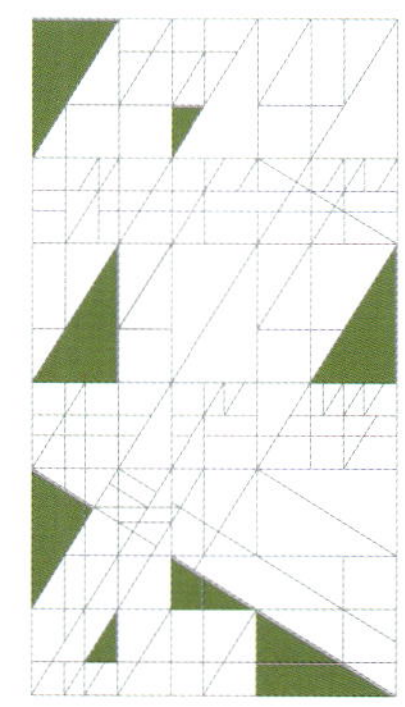

Cespugli e siepi

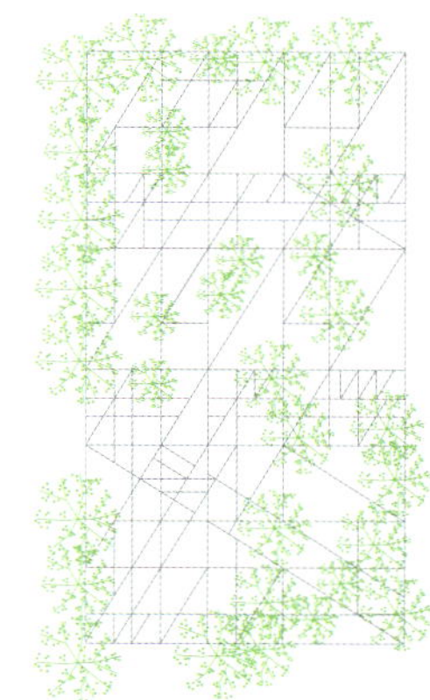

Alberi

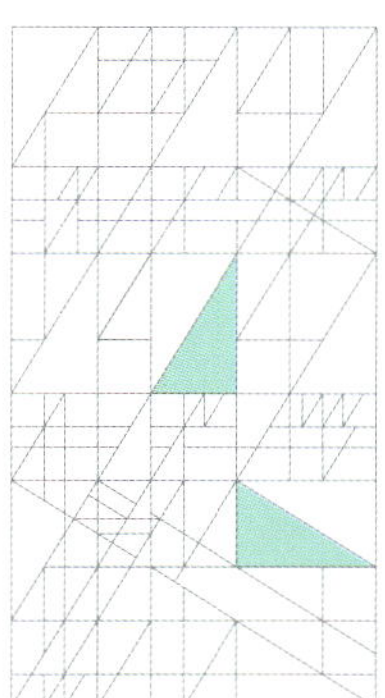

Acqua

Pietra scura

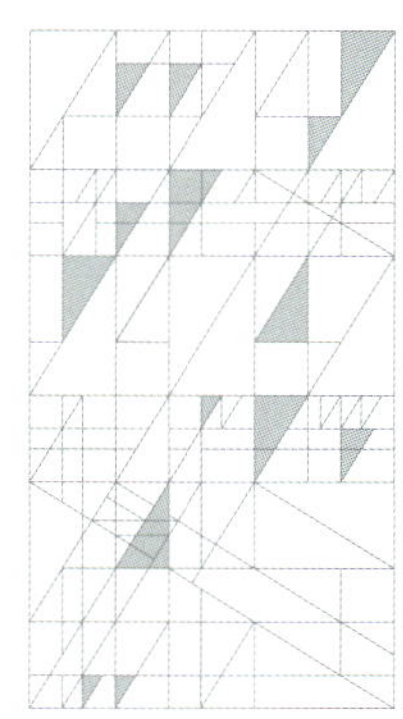

Pietra chiara

Cemento

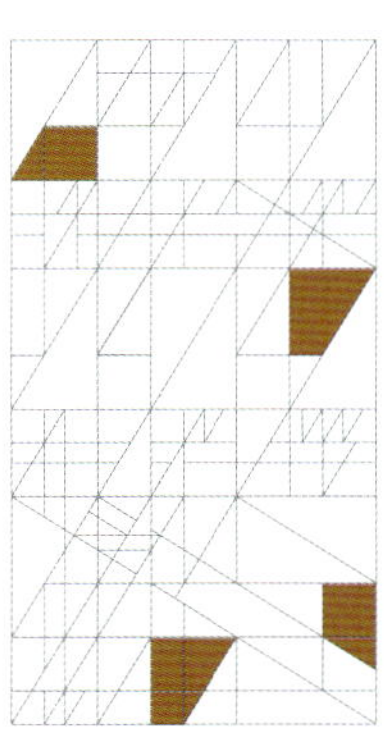

Legno

Panchine

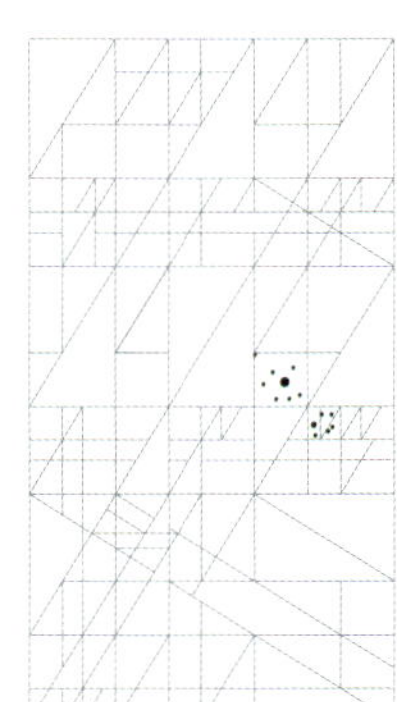

Giochi

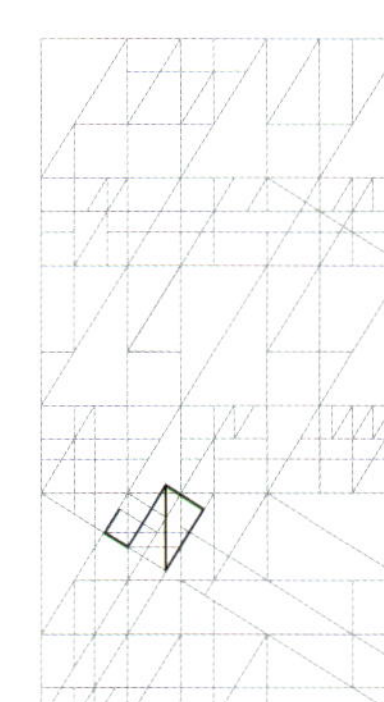

Porta Europa

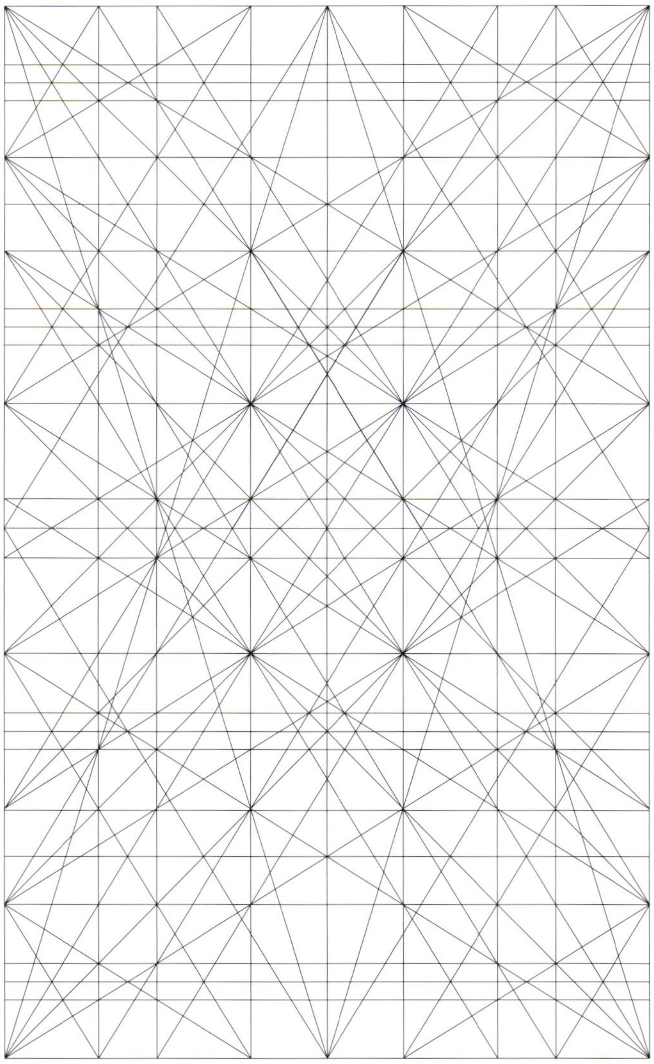

Detail n°12 of Porta Europa

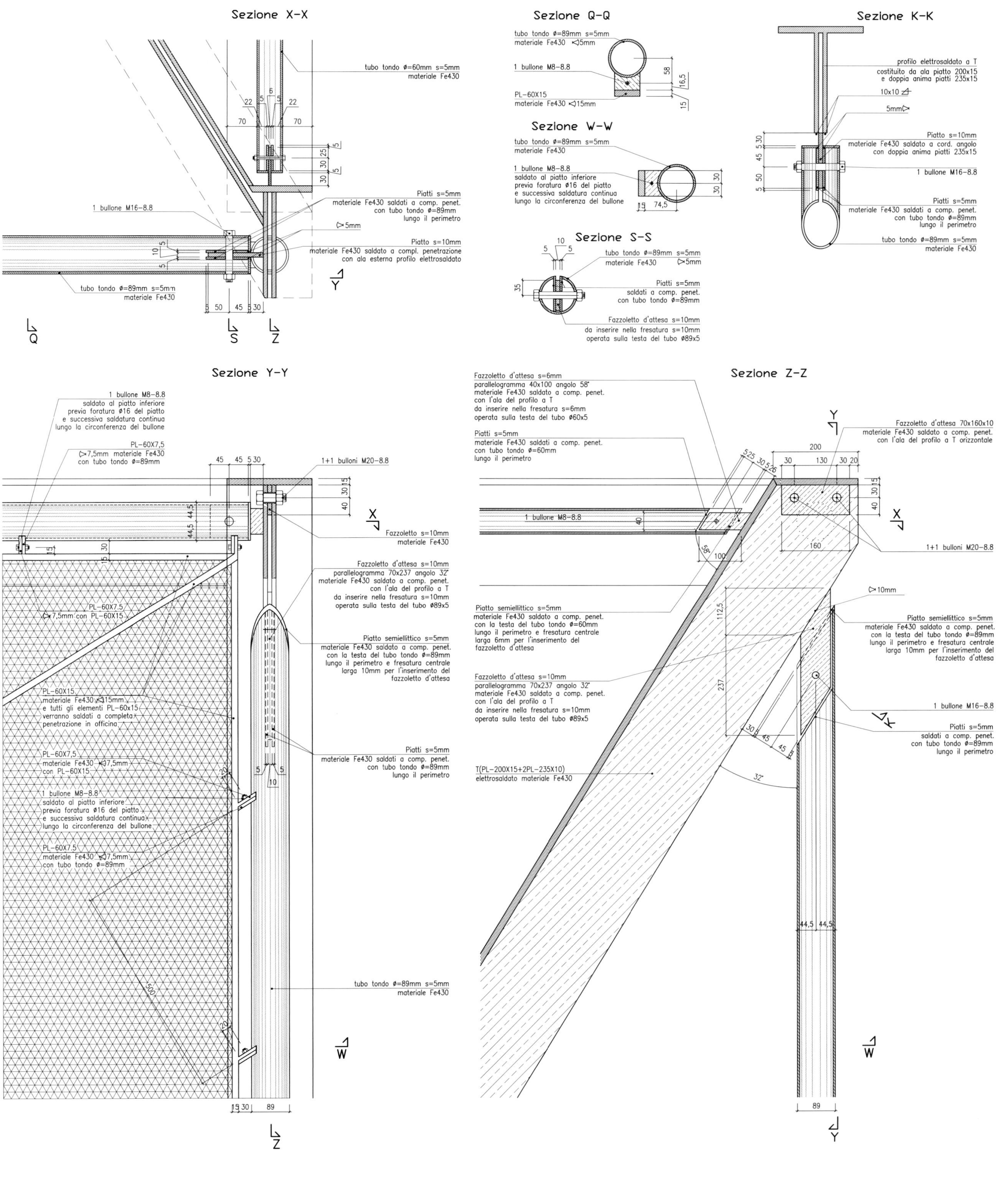

BRADFORD'S CITY PARK

布拉德福德城市公园

Bradford, Yorkshire, UK

The completed design has created a flexible dynamic centerpiece in the form of a vibrant 24,000 square meters public space, including the mirror pool, fountains and public art. The park centers on Bradford's 19th Century City Hall and helps to attract the city's major visitors' attention and to connect with transport hubs and the rest of the city center. It enhances the overall image of Bradford and helps create a landscape for investment by setting Bradford apart from other cities.

The development and implementation of City Park's design by Gillespies from the original masterplan has been founded on three fundamental design concepts:

1.HINTERLAND: Bradford's setting is that of a city surrounded by visible hills and countryside. The design for City Park seeks to suggest the meeting of urban and rural. Gillespies wanted the idea of hinterland to reference the outside towns and villages looking into the City, and the City looking outwards. This concept is borne out not only by the park's design and built forms, but also how City Park interacts with Bradford's people – both those within the city, and those on the edges attracted into the city by its regeneration.

2.WATER: The unifying element of City Park's design is water. Water provides a deep-rooted connection to Bradford's industrial energy.

3.MIRROR: City Park's space provides a place to reflect. Its mirror pool provides an architectural mirror to Bradford's city center, sky and weather. The mirror pool brings this public space alive and provides a mirror for events, people and culture.

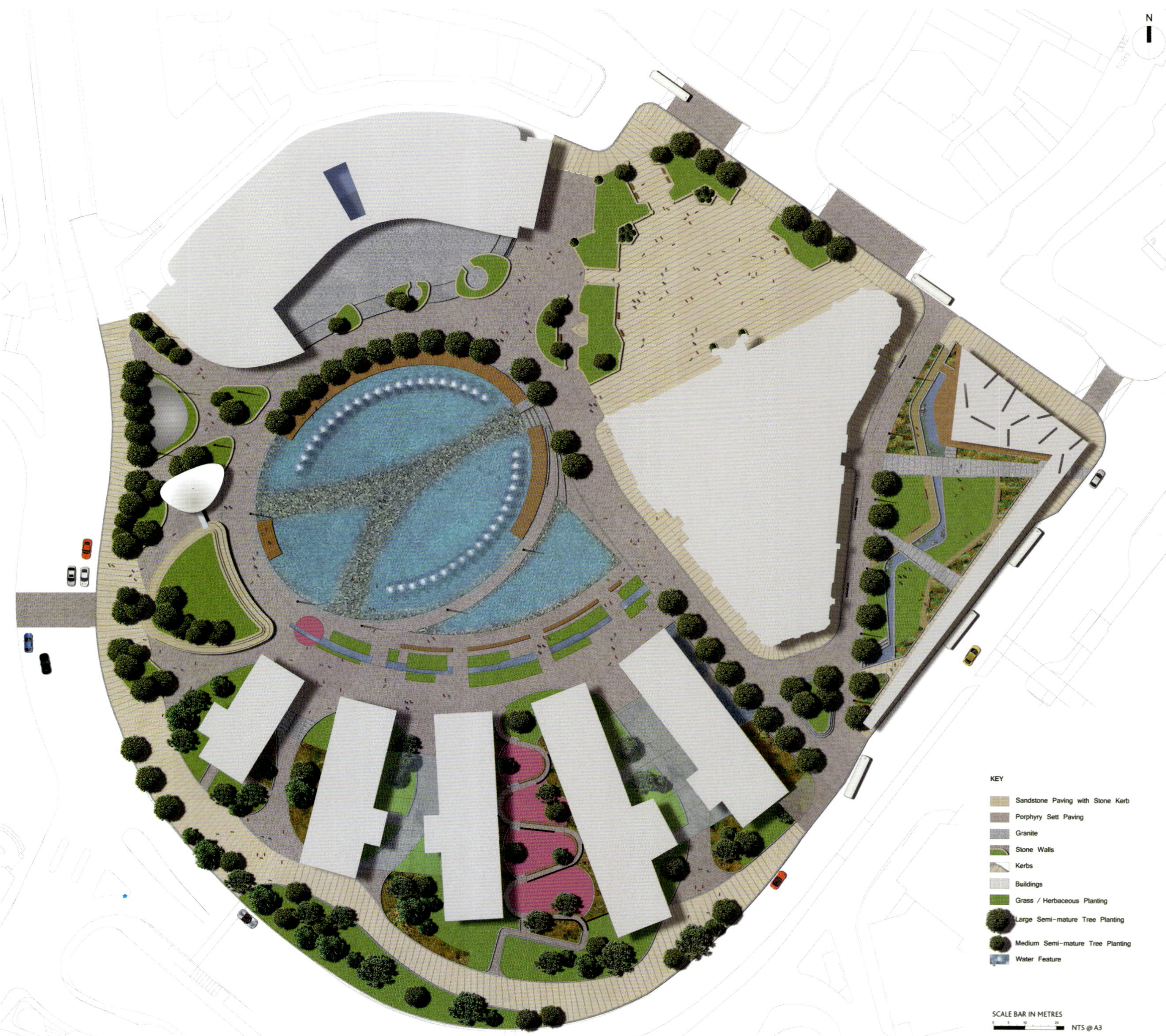

- **Area** - 24,000 m^2
- **Photographer** - Giles Rocholl

Football pitches

Piccadilly Gardens (Manchester)

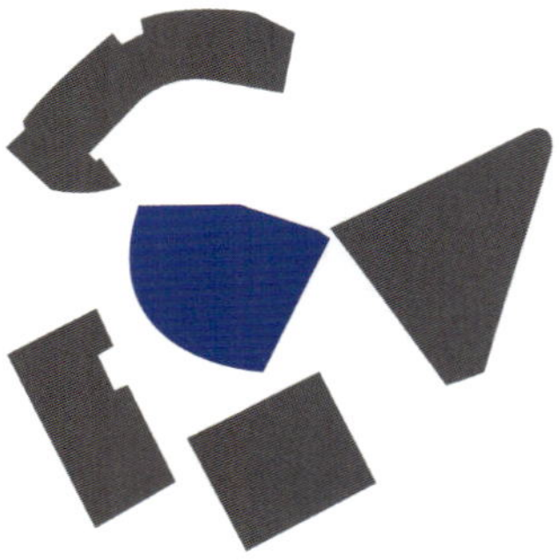

Peace Gardens (Sheffield)

St Marks Square (Venice)

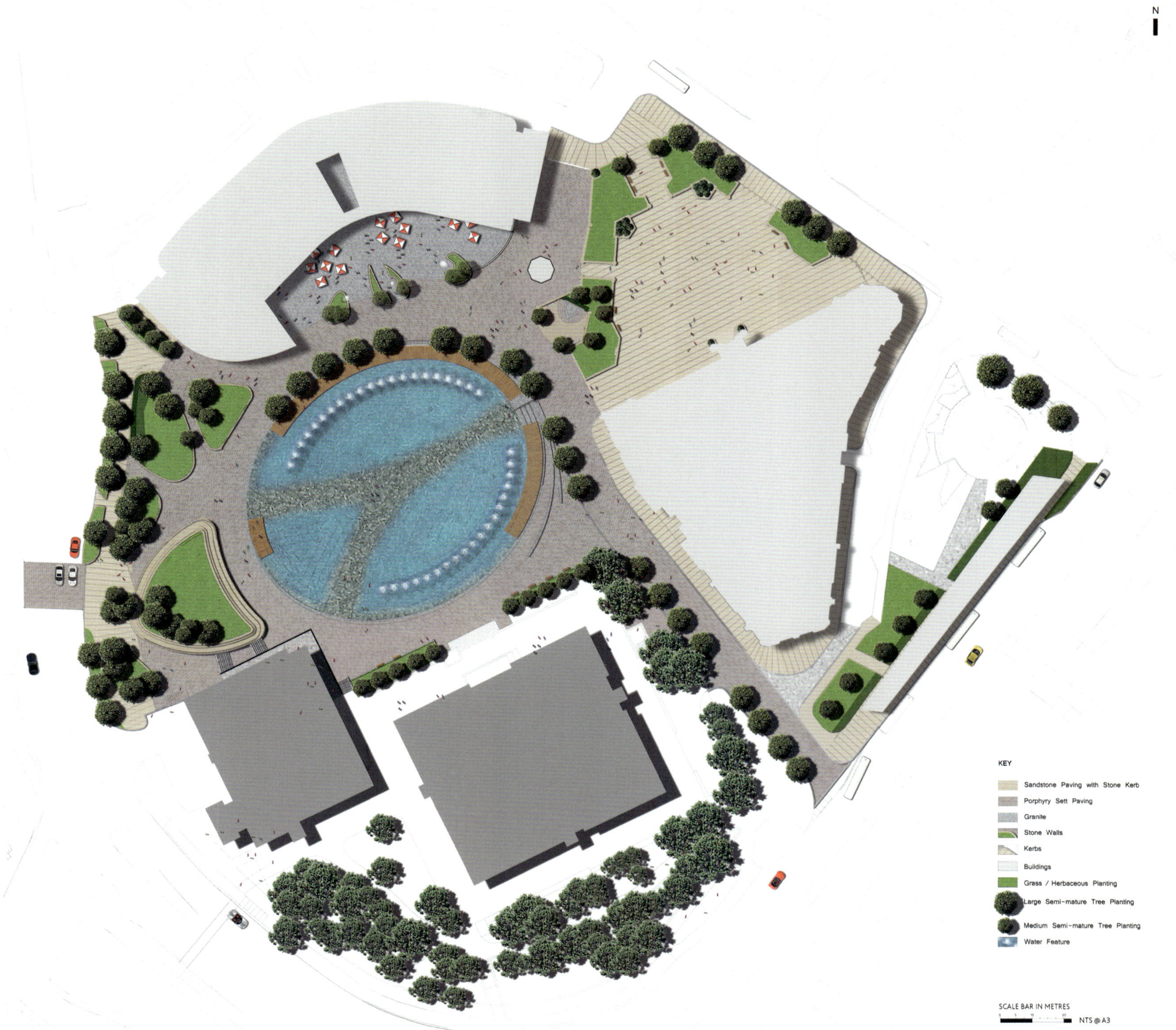
N
KEY
Sandstone Paving with Stone Kerb
Porphyry Sett Paving
Granite
Stone Walls
Kerbs
Buildings
Grass / Herbaceous Planting
Large Semi-mature Tree Planting
Medium Semi-mature Tree Planting
Water Feature
SCALE BAR IN METRES
NTS @ A3

MODERN OFFICES TO LET 01274 759955

citypark
CITYPARK

CENTENARY SQUARE

布拉德福德城市公园占地约24000平方米，整个风格活泼动感，空间布置灵活多变，包括一个如镜般的池塘、喷泉和公共艺术作品。公园位于19世纪的布拉德福德市政厅的中心，这样的设计布局可以吸引游客，同时，可与城市的公交枢纽以及其他地区相互连通，便于市民出行。城市公园的建成为布拉德福德市区增添了新的景色，也为市政投资做出了极大的贡献。

吉尔斯皮斯建筑事务所全权负责整个城市公园的设计规划进展及施工，最初的设计规划基于三个主要的设计理念：

1. 山景

布拉德福德市位于山区，周围群山环绕。城市公园的设计力求将城市化与乡村风格结合。吉尔斯皮斯建筑事务所提出了山景的概念：以周边的乡村和城镇为参照物，与城市化概念形成鲜明的对比和差异。山景的概念不但体现在城市公园的设计和建筑模式上，更重要的是体现城市公园与布拉德福德市的民众之间的互动。山景的概念不仅造福于居住在市内的居民，还会使居住在城市边缘的居民对于城市的重建产生一种认同感。

2. 水景

水景将城市公园的各个元素融合为一个整体。它将布拉德福德市的工业能源紧密地联系在一起，形成一个根深蒂固的体系。

3. 镜像

城市公园占地广阔，宽阔的池塘表面如同镜子般将建筑景观、城市中心、天空以及美景倒映在水中，形成美丽的镜像，为整个城市公共空间带来了一股动感的气息，也反映了当地民众和当地文化的独特风格。

MUSIC PARK IN SEVILLE

Seville, Spain

塞维利亚音乐广场

The Plot is located between Barrio de Nuestra Señora del Águila and Barrio de la Música in Seville.

Parque de la Música is a unique masterplan in terms of urban renovation and integration linked to Seville Underground infrastructure works. The Park area is 27,232 m^2 and ensures accessibility to Cocheras Station, located at the center of the Park, and the connectivity between the houses located at both sides of the station.

The new landscape helps people enjoy a walk around it or have a rest in the designated areas of the park, and for the continuity of different urban itineraries, make them more convenient and accessible for everybody.

The park is organized along a north-south promenade that reaches the station level with a gentle slope. Other transversal circulations connect the main promenade with the surrounding streets and plazas.

In the east side, the difference of levels is solved with retaining walls embracing and protecting the plazas at the surrounding neighborhoods' levels. There is a lift and an escalator next to the station entrance for accessibility.

In the west side, the level difference is resolved with a retaining wall that goes parallel to the underground tunnel and curves itself creating the ramp directing to the main west entrance. There is another lift for access at the corner between the pedestrian street Águila Marina and Águila de Oro Street.

First quality pavement materials have been used: ceramic tiles, cobblestone of lime and granite stones. The patterns are repeated creating a geometry that connects all the different areas in the park. This pattern is inspired in an existing ceramic pattern in the Patio de Muñecas from the Reales Alcazares in Seville.

There is a dual purpose: a more cohesive city, regenerated and with better internal connections, and providing a more human and better quality outdoor space and therefore better quality of life for their inhabitants.

• **Gardening and Landscaping** - ERNESTO FERNÁNDEZ SANMARTÍN, SARA TAVARES COSTA

• **Area** - 27,232 m^2

• **Quantity Surveyor** - Antonio Carrascal Cruzado

• **Photographer** - PABLO DÍAZ – FIERROS

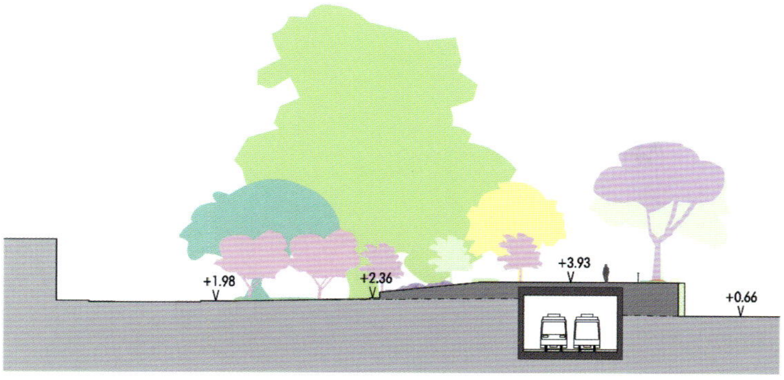

SECTION 1

SECTION 2

SECTION 3

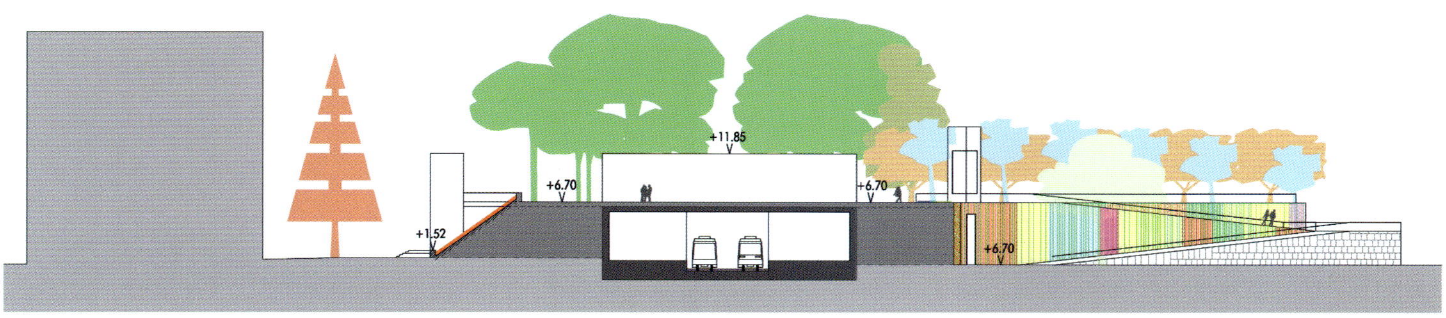

SECTION 4

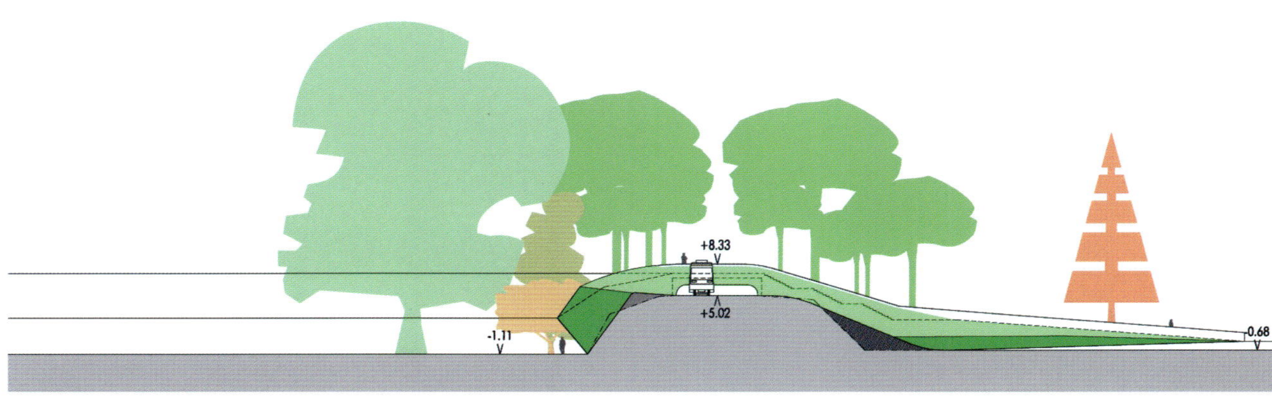

SECTION 5

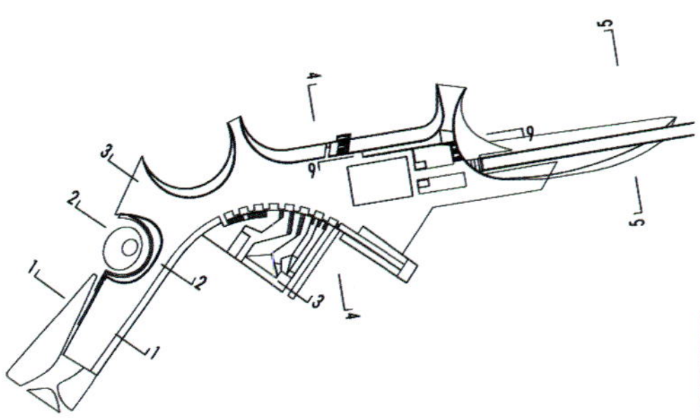

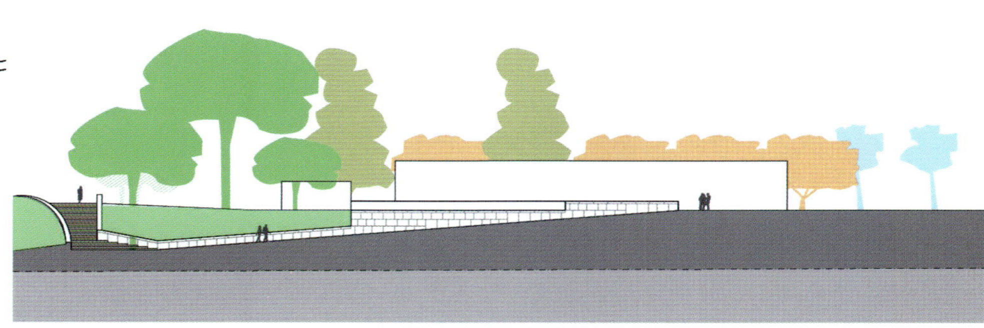
SECTION 6

塞维利亚音乐广场位于塞维利亚市区的两个居民区之间。

塞维利亚音乐广场工程是整个城市重建工程中最重要的环节之一，广场与塞维利亚的地下基础设施工程相连。广场占地 27232 平方米，地理位置优越，便于市民前往广场中央的科奇拉车站，在车站至广场两边的建筑之间都设立了专门的通道。

新的广场景观为在广场漫步的人们带来了更多的视觉享受。设计师们在广场上专门设置了休息区，尽可能地为不同路线的游客在市区游览提供便利。

整个广场沿着南北走向设计了供行人散步的主通道，这样人们可以在较小的坡度上步行至车站。另外，横向分布的环形步道连通了人行主通道和周边的街道以及露天广场。

广场东部保留了原有的围墙，不但起到了防护周边露天广场的作用，而且在视觉上自然消除了东西两侧的水平落差。靠近车站的一侧，还设计了电梯和自动扶梯，方便公众进出车站。

广场西侧，同样保留了部分围墙用以消除水平落差，围墙呈弧形，并与地下隧道平行，在西侧入口处形成一道斜坡。另外，一部电梯设在玛丽娜街与奥诺街的交汇处。

设计师们采用了最高级的建筑材料来铺设步行通道：瓷砖、鹅卵石、花岗岩石板等。设计师从塞维利亚的阿卡莎宫的园林设计中得到灵感，将这些材料组合成重复的单元顺着通道延伸，连接广场的各个部分。

广场的设计具有双重意义：一是使重建后的广场具有更好的内部环境，从而提升城市的凝聚力；二是为市民提供一个更为人性化、更高质量的户外空间，提高市民的生活质量。

• **Architect** - Javier Fresneda Puerto & Javier Sanjuán Calle (MTM architects), Javier Peña Galiano (Xpiral architect) • **Lighting** - Mario Gentile

PORMETXETA SQUARE

PORMETXETA 广场

Bilbao, Spain

The Animated Square

The Pormetxeta Square is founded under the long ramps. There are some holes that qualify the space with playgrounds and lend it a unique topography. Big artificial trees of 11.5 m tall, made of COR-TEN steel, hang boulders on the cup of the tree, and work as a parasol. Beneath them, all you can hear are the laughter of children, the roller skaters, people's talking…

The Pormetxeta Square and its different levels of access make up a unique and independent project, whose contrasts intensify its purpose.

Equipped Communications Arteries

The very steep slope of the plot was scan as an opportunity for the creation of a new urban typology. It was possible to take advantage of the 20-meter difference between the upper and bottom level of the plot, to build an equipped slope, "the Pormetxeta Square", with the following strategies:

First of all, several accesses and approaches are linked: the railway platform, the highway connection, the entry towards the promenade river, the pedestrian and road streets, the access from the high and low street, the Desert Square, the school …

Secondly, they saved all the space under each "Pormetxeta Ramp" in order to get 2,341 m^2 of commercial and service public facility.

The third advantage is working as a micro-public space: micro-squares, meeting points, rest stands, outdoor stays, in other words building a phenomenological sequence.

Finally, it acts as a multifunctional element. It works as street furniture, balustrade, pergola, lantern… creating a very well-equipped place.

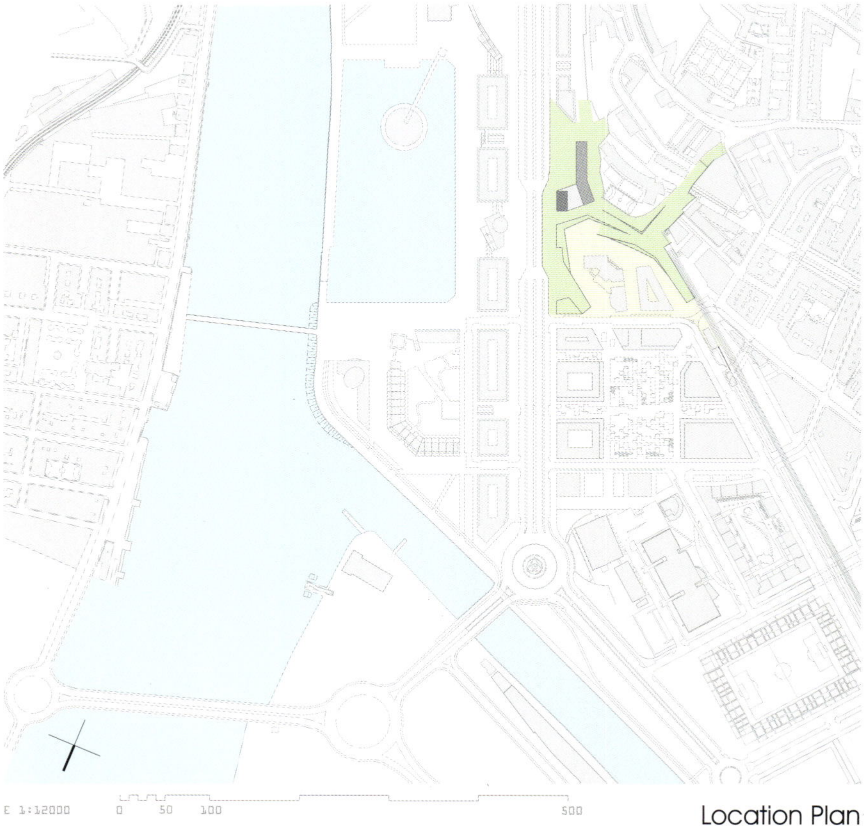

Location Plan

• **Structural Consultant** - I.D.E.E.E. Eduardo Diez Gernier

• **Technical Assistant** - Dair ingenieros, Jose Ángel Laguna, Enrique Álvaro

• **Photographer** - David Frutos | Bis Images

Floor of Decks

First Floor

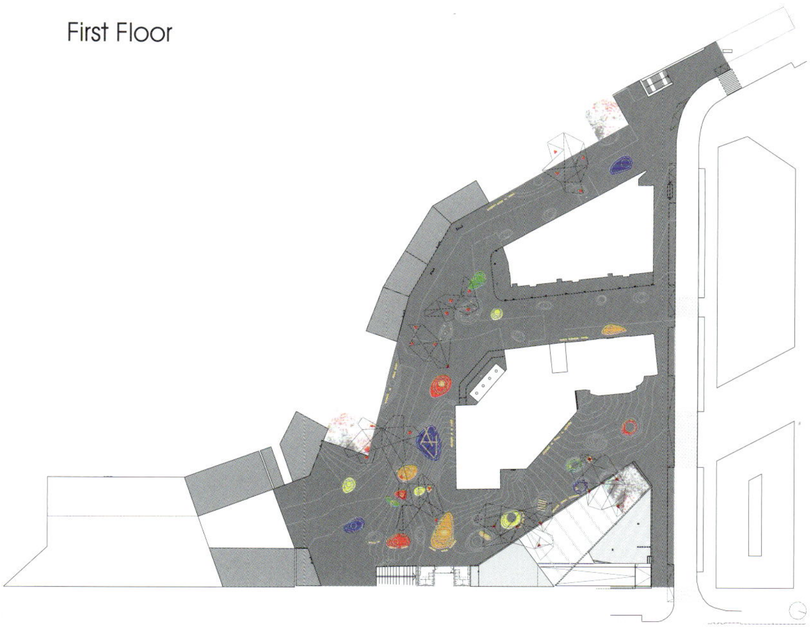

Second Floor

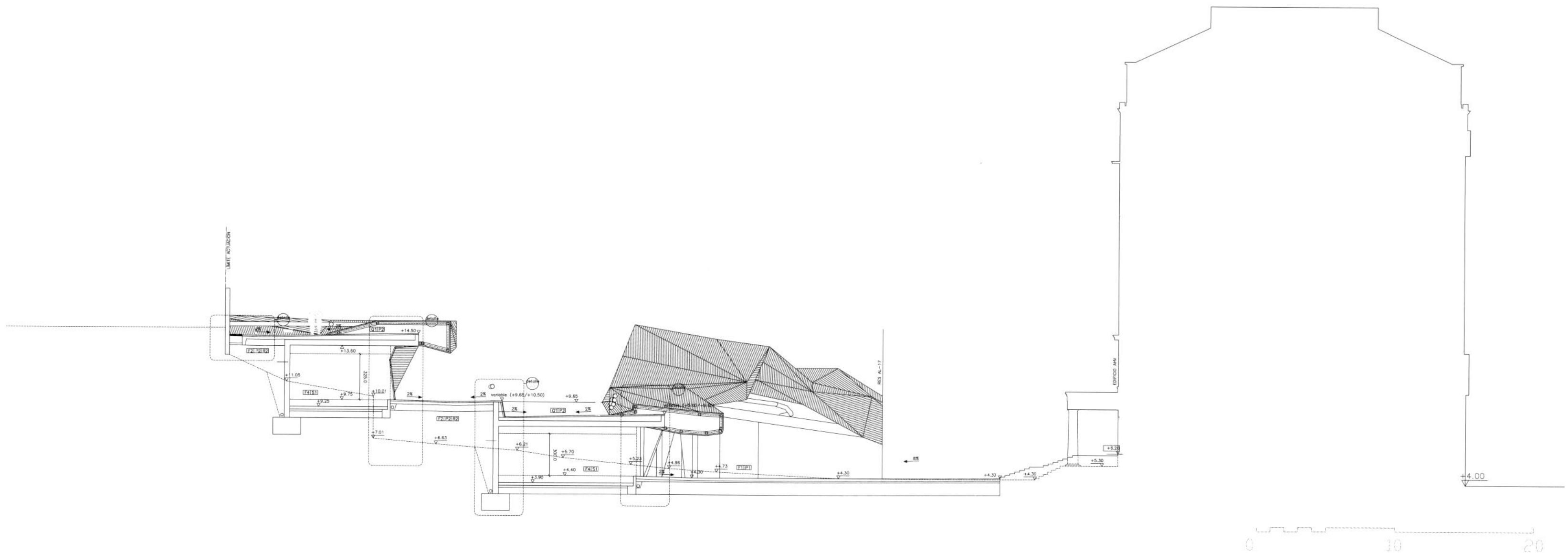

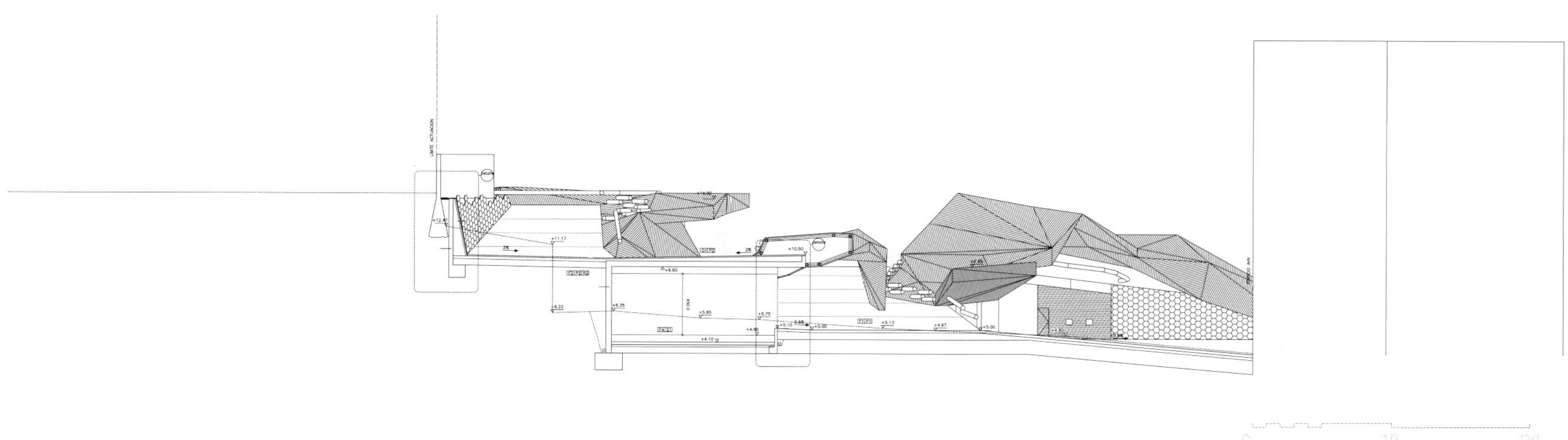

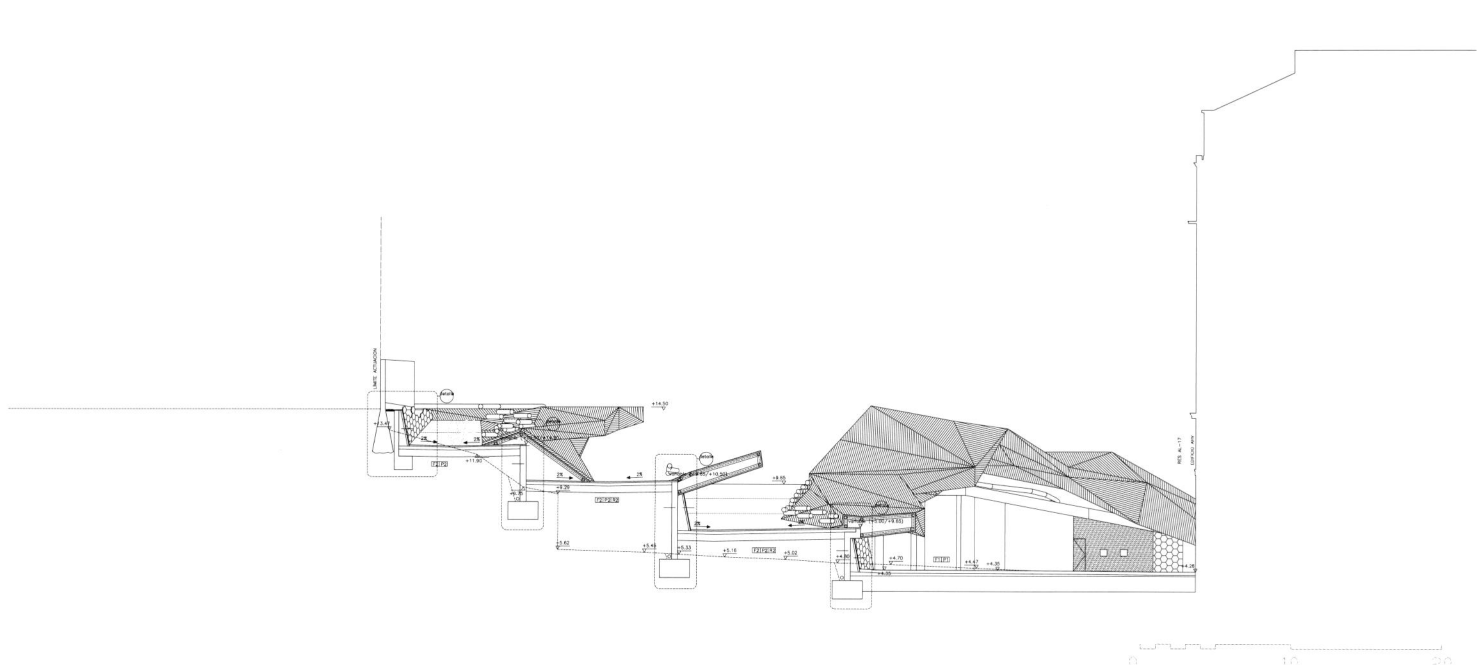

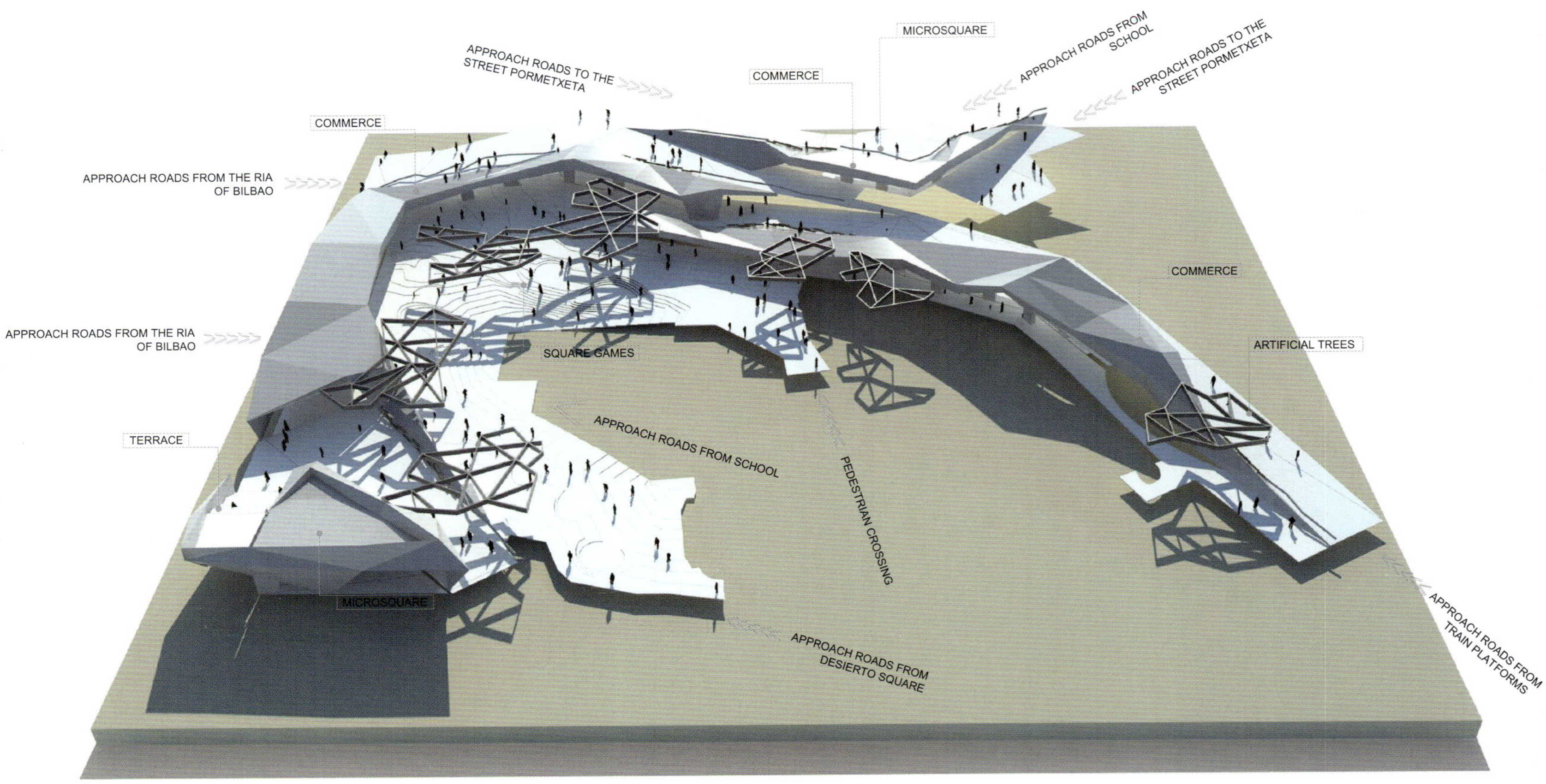

Render By Xpiral

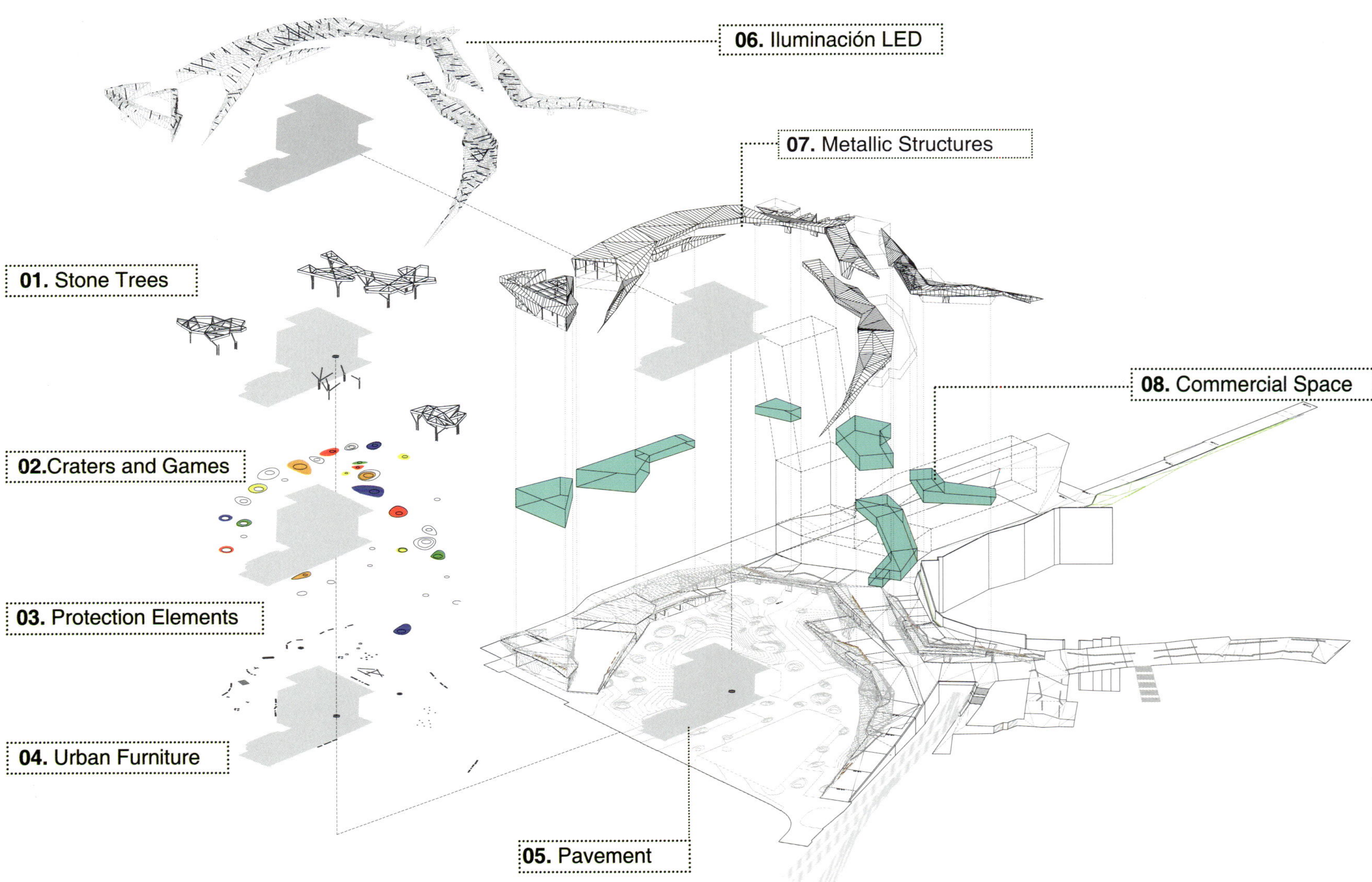

Axonometric Of The Complex

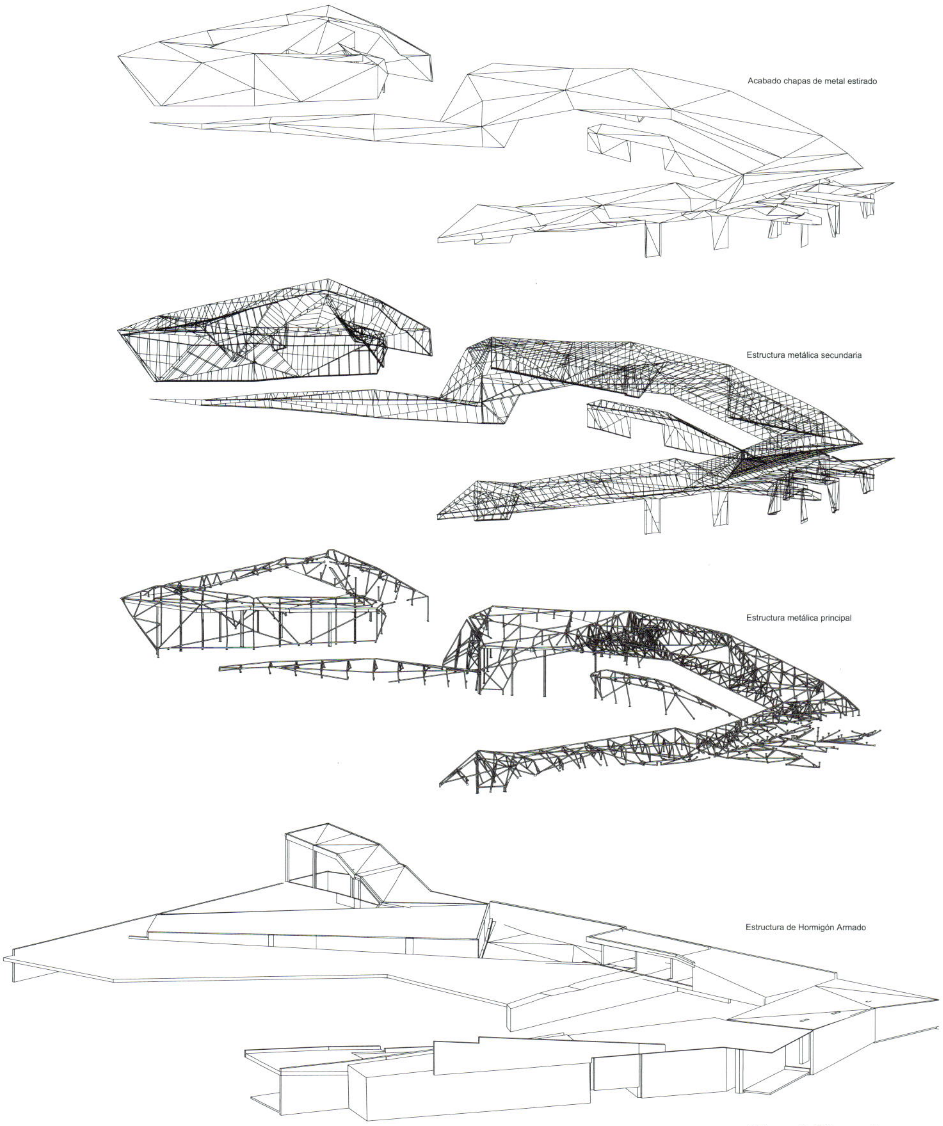

Steel Structure

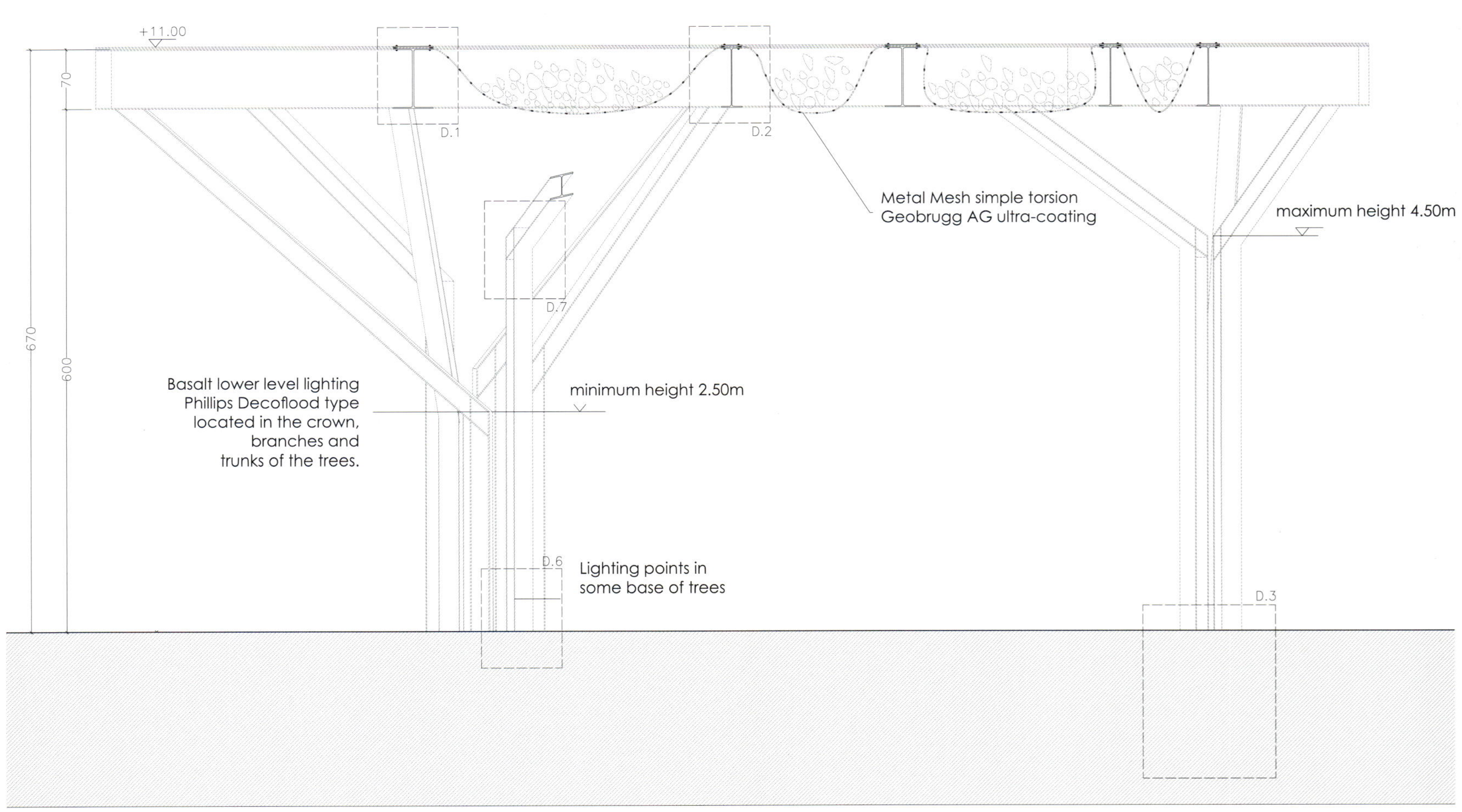

+11.00
70
670
600
D.1
D.2
D.7
D.6
D.3
Metal Mesh simple torsion
Geobrugg AG ultra-coating
maximum height 4.50m
Basalt lower level lighting
Phillips Decoflood type
located in the crown,
branches and
trunks of the trees.
minimum height 2.50m
Lighting points in
some base of trees

Stone Trees

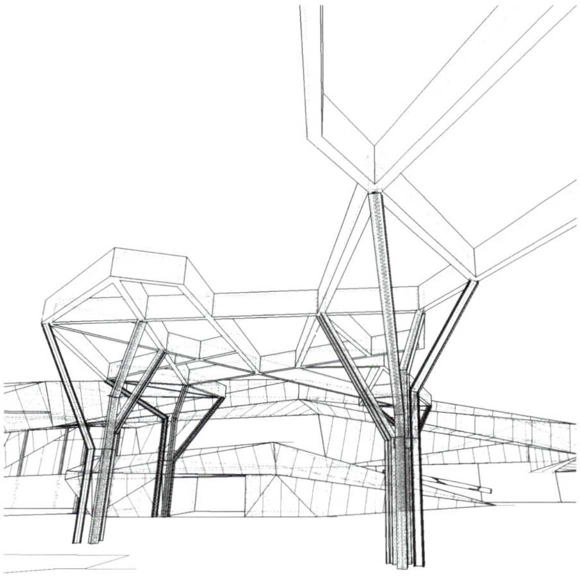

The stone trees are hybrid elements that amalgamate a number of functions, ranging from infrastructure, shaping the urban landscape, to an atmospheric device that balances the exuberant natural surroundings.

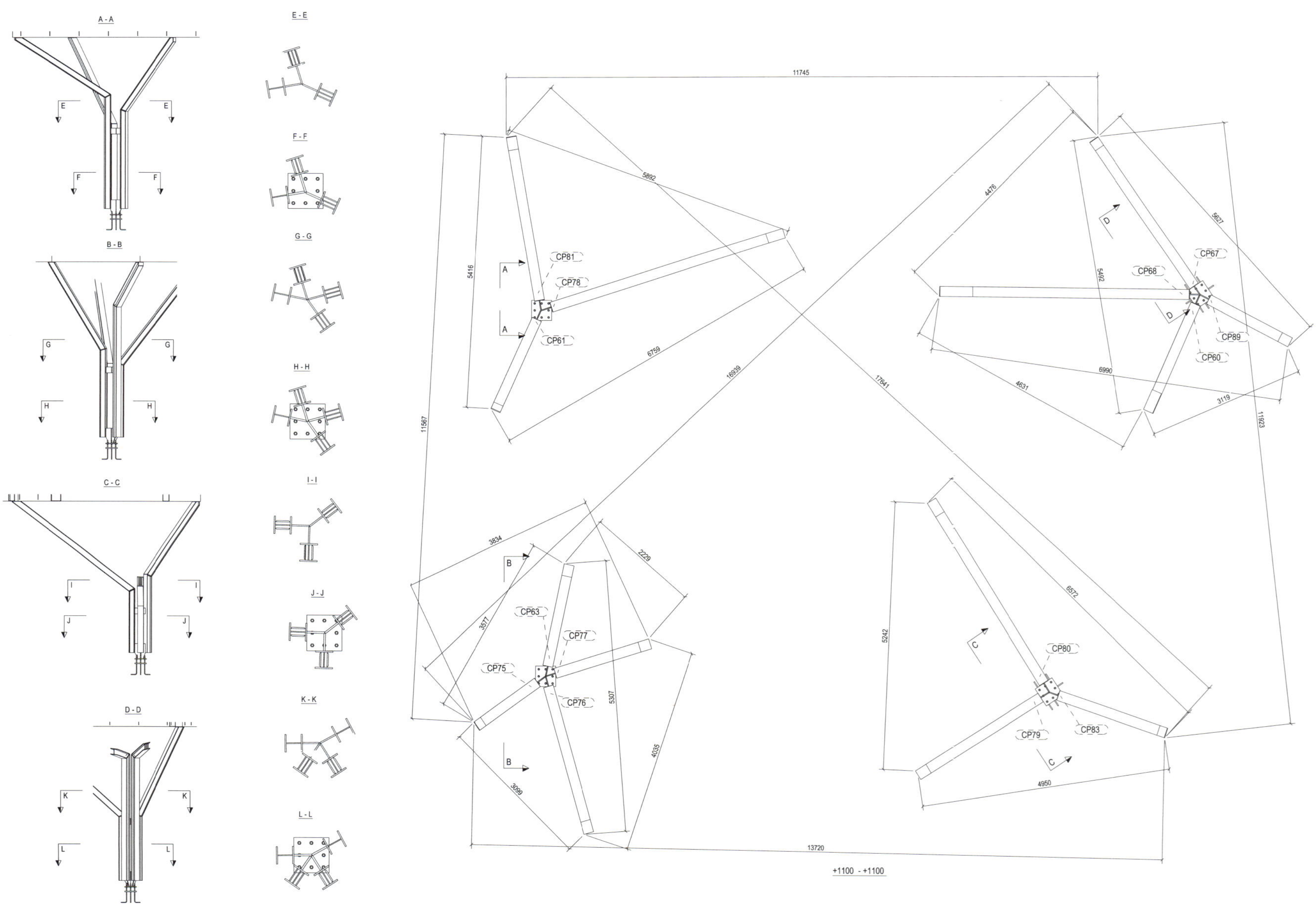
A - A
B - B
C - C
D - D
E - E
F - F
G - G
H - H
I - I
J - J
K - K
L - L
CP81
CP78
CP61
CP68
CP67
CP89
CP60
CP63
CP77
CP75
CP76
CP80
CP79
CP83
11745
13720
+1100 - +1100

Stone tree - type 1

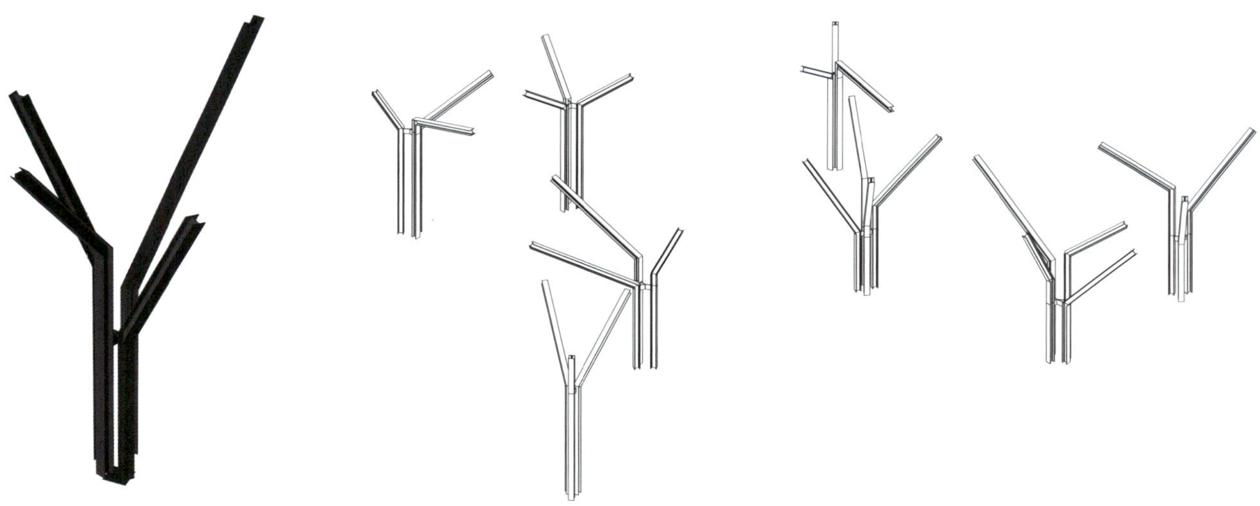

Stone tree - type 2

充满生气的广场

西班牙的 Pormetxeta 广场建在一个长坡道上。广场里有些洞，这些洞构成了几个运动场，增添了其独特的地势地貌。高达 11.5 米、由科尔坦耐大气腐蚀高强度钢制作而成的人工大树垂挂着遮住树下的卵石，如一把大伞。在树下可以听到孩子们的笑声，滑旱冰的声音以及人们的谈笑声等等。

Pormetxeta 广场及其不同层次的通道构成了一个个独特的工程，各有千秋，各司其职。

装备完善的通讯动脉

该地块陡峭的斜坡地形被视为创造新城模式的绝好机会。设计师充分利用了该地块 20 米的高度差，精心设计建成了 “Pormetxeta 广场”，以下的策略为此提供了可能性：

首先，所有的通道是相互连接的：车站月台、公路连接、通向海滨人行道的入口、步行街、主次大街的通道、沙漠广场和学校之间的通道等等。

其次，为了保证 2341 平方米的商业和公共设施服务区面积，节省了每一块“Pormetxeta 斜坡”下的空间。

再次，该广场可以作为一个微公共空间：微广场、聚会点、休息站、户外停留站，换言之，广场建立起一个现象学顺序。

最后，广场具有多种功能，有街道设施、栏杆、藤架、灯笼等，是一个装备精良之地。

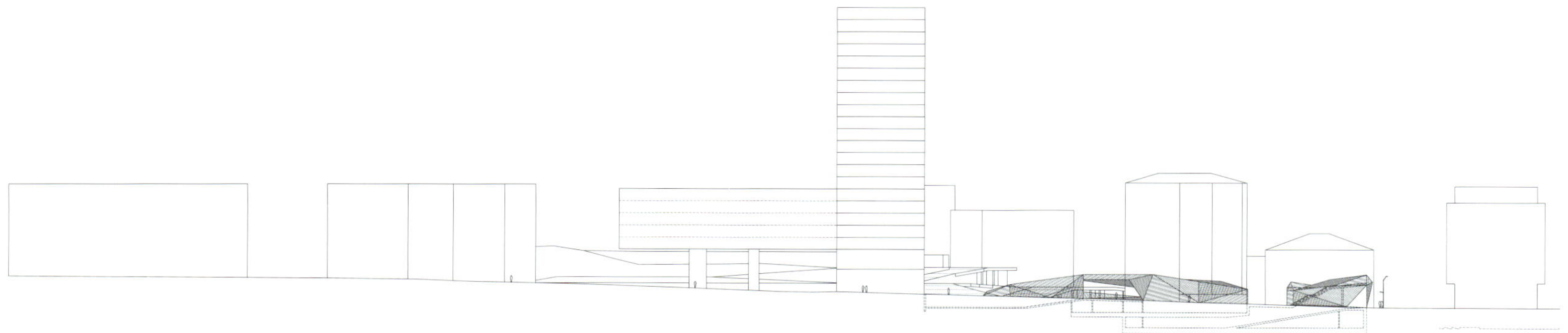

• **Landscape Architect** - Miralles Tagliabue EMBT • **Area** - 9,200 m² • **Photographer** - ALEX GAULTIER

PLAZA RICARD VIÑES

理查德广场

Lleida, Spain

The project is made up of a circle and is a new square for the city. The circle ends in a labyrinthine pattern, made up of bands of planting and brick paving.

The square is split up by green areas; a maze of trees and low shrubs that are visible from the roadway. This greenery helps organize the neighborhood's leisure activities with bars, children's play areas, benches and pedestrian pathways. The two streets that feed into the square will be pedestrianized, using natural stone paving, and they will be softened by new tree planting and small green spaces which will punctuate the entrances to residential buildings. From one side of the circle a wide sculptural light source will illuminate the whole square, underlining both the idea of the community that inhabits a public space and the street party concept as a mark of collective identity. A new sculpture installed in the middle of the circle will become the landmark that makes this place stand out from the rest of the city.

GROUND FLOOR PLAN

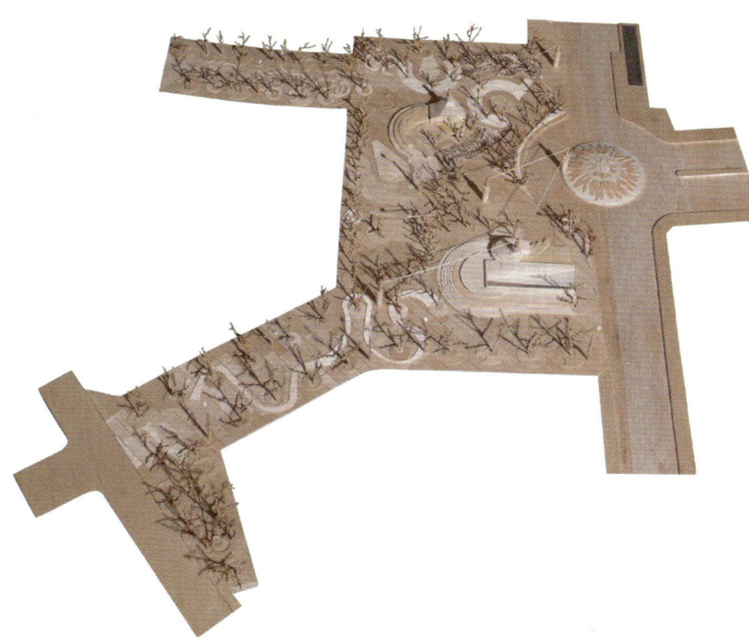

理查德广场项目是一个整体呈圆形的新建广场，广场内的规划形式如同一个迷宫，由多条绿化带和砖砌小路构成。

广场内部由绿化带和绿地分割为不同的区域，在广场的路边就能看到一排排高大的树木和低矮的灌木丛。这些绿地和草木，配合酒吧、儿童乐园、长椅以及人行小路，为周边民众构建了一个休闲活动的场所。两条贯穿广场的主通道由天然的石块铺成，作为行人专用通道，两侧种植了绿树和青草，缓和了石块路面带来的突兀感和粗犷感，同时，在周边建筑的背景下，有效地突出了公园大门的视觉效果。广场的一侧设有大片雕塑造型的灯光系统，入夜后灯光开启，整个广场如同白昼。这里不仅是一个普通的公共活动区域，也是一个街区社团集体的象征。圆形广场的中心矗立着一尊崭新的标志性雕塑，象征性地突出了理查德广场在城市中的社会地位。

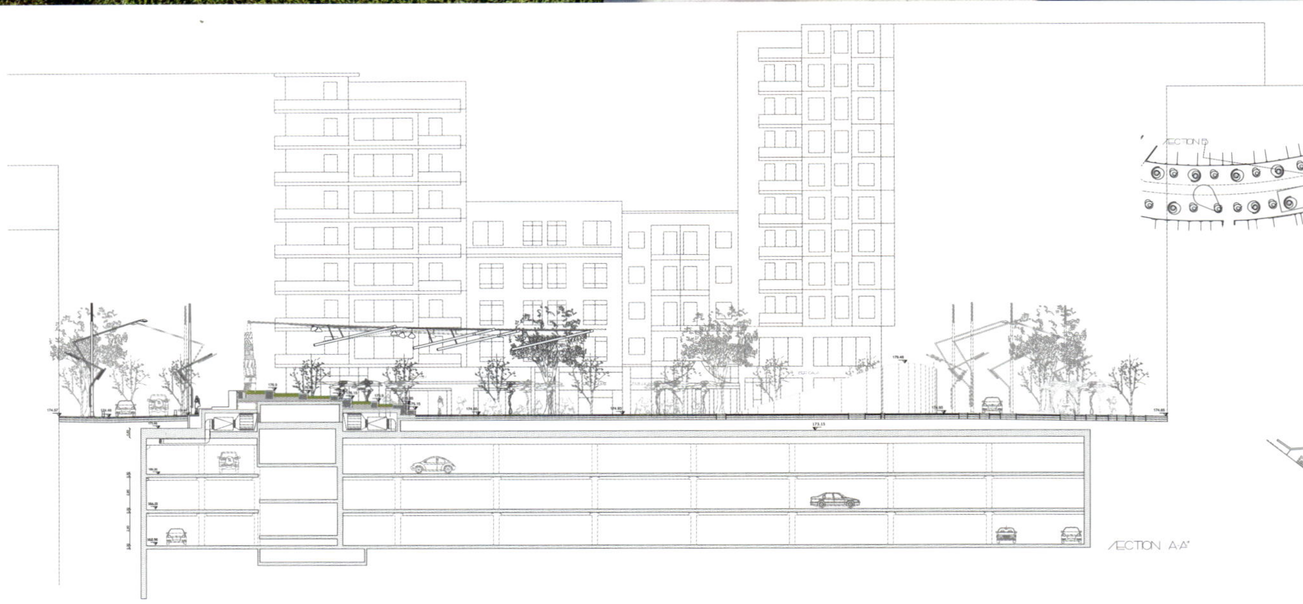
SECTION A-A'

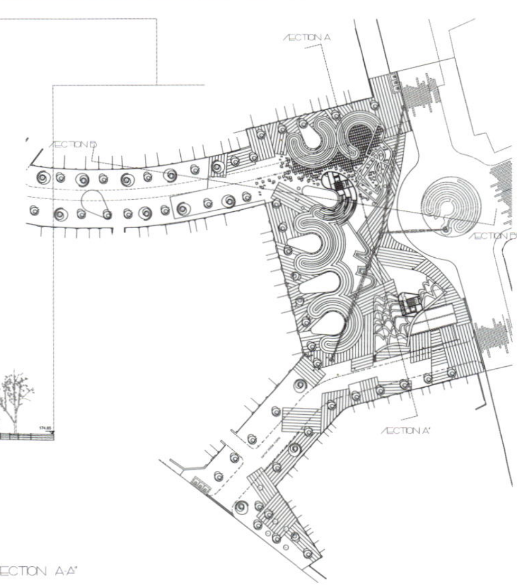
SECTION A
SECTION B
SECTION B'
SECTION A'

SECTION B-B'

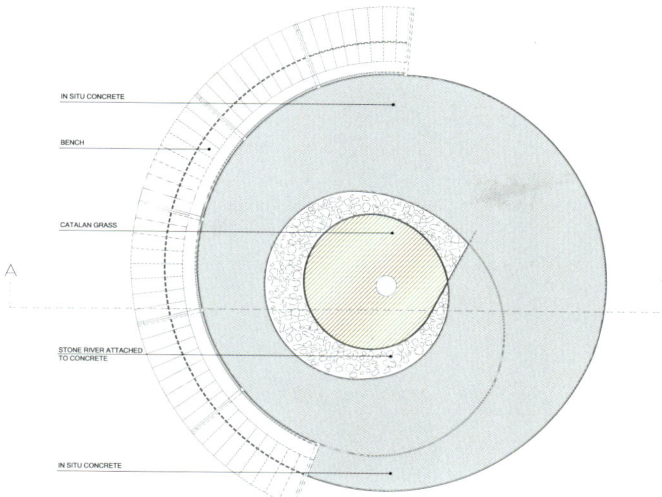

TREE GRATING AND BENCH PLANT

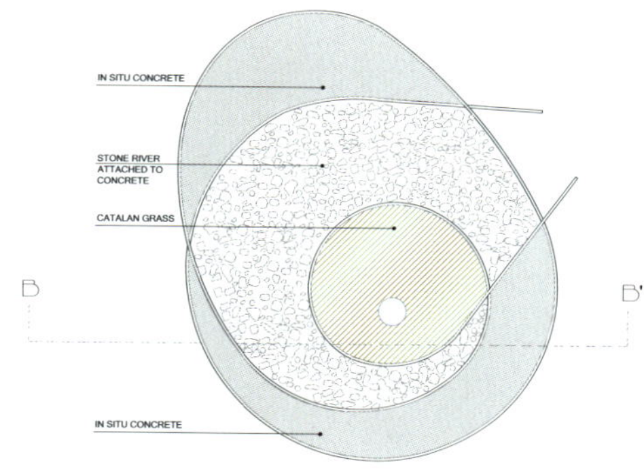

TREE GRATING PLANT

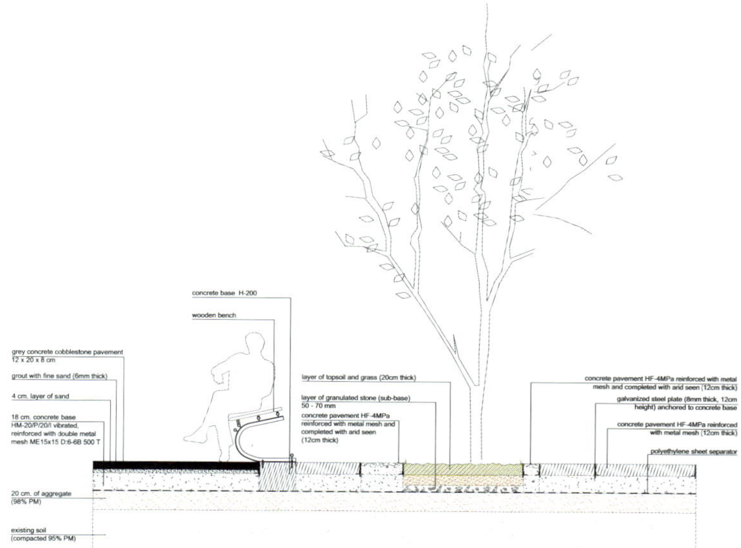

SECTION A - A'

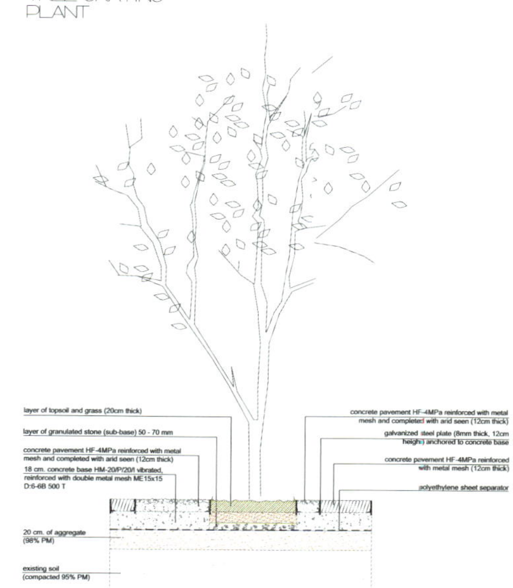

SECTION B - B'

BENCH

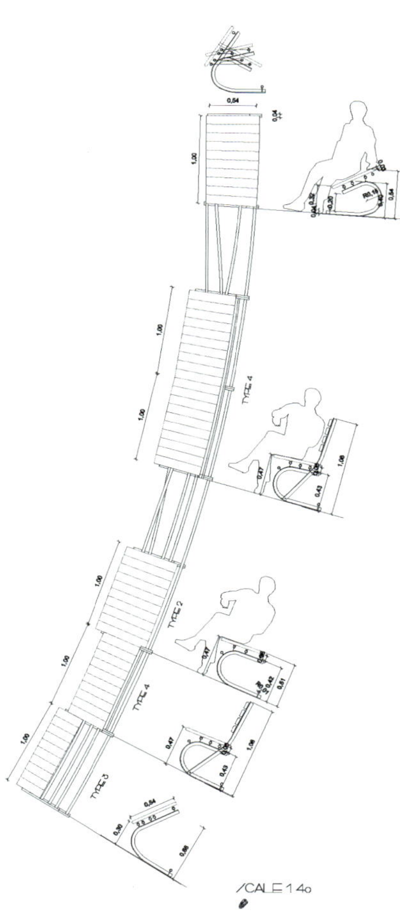

SCALE 1:40

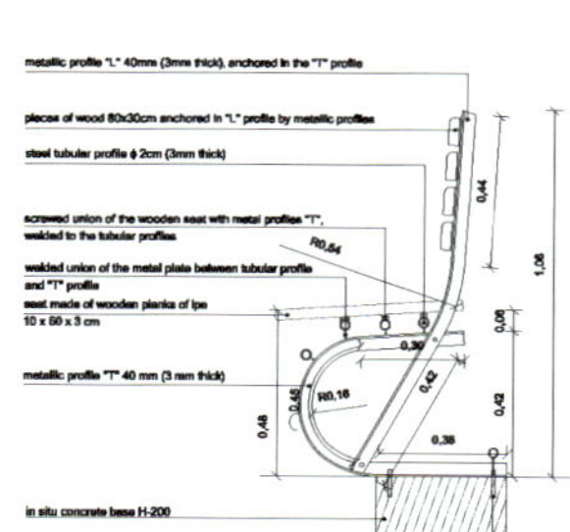

SCALE 1:20

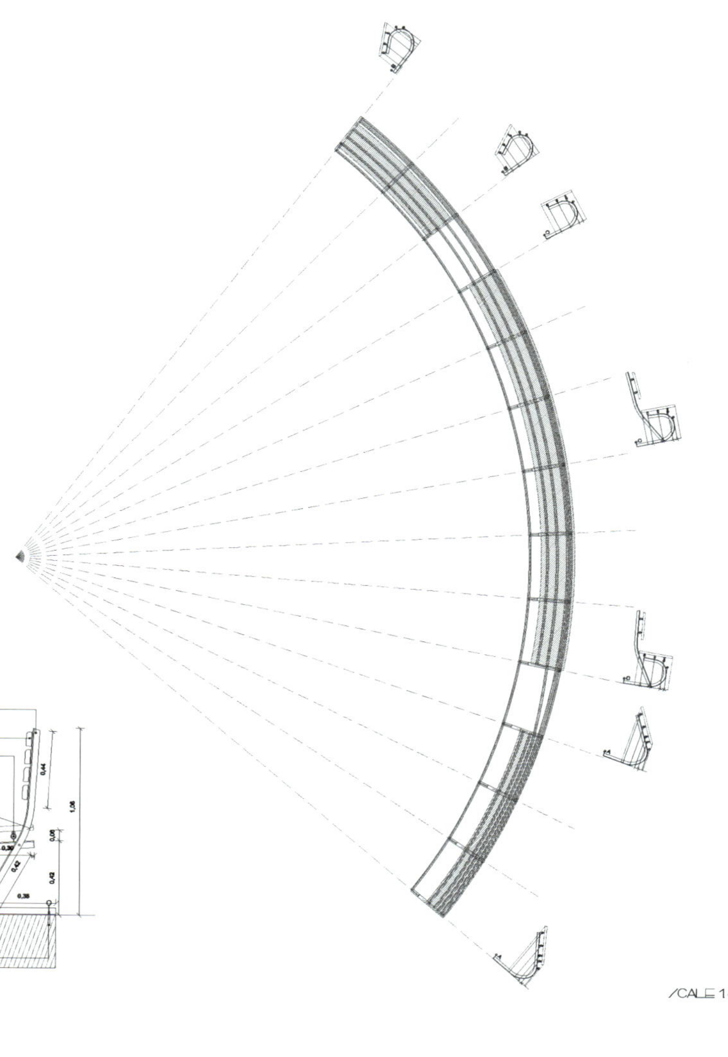

SCALE 1:50

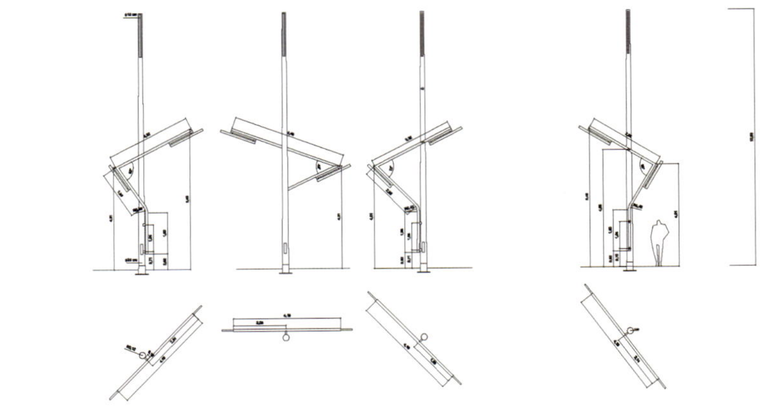

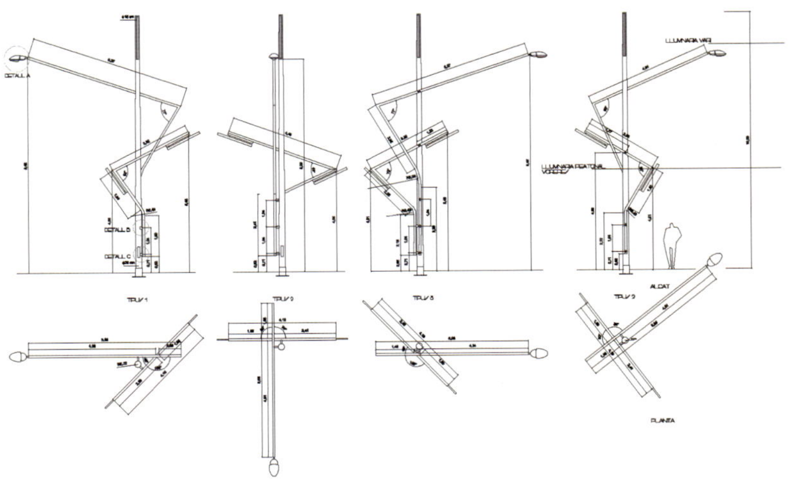

STREET LIGHTNING

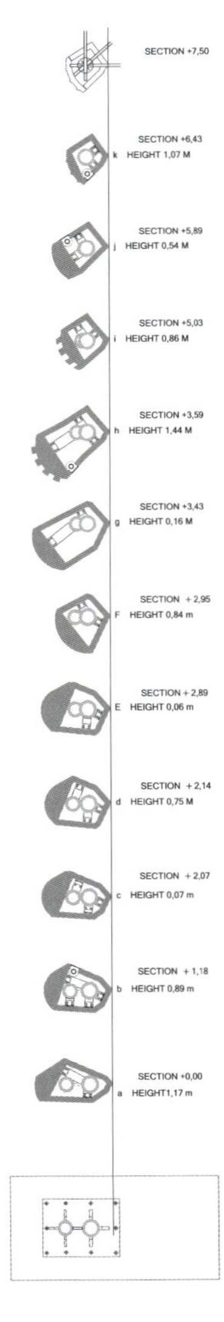

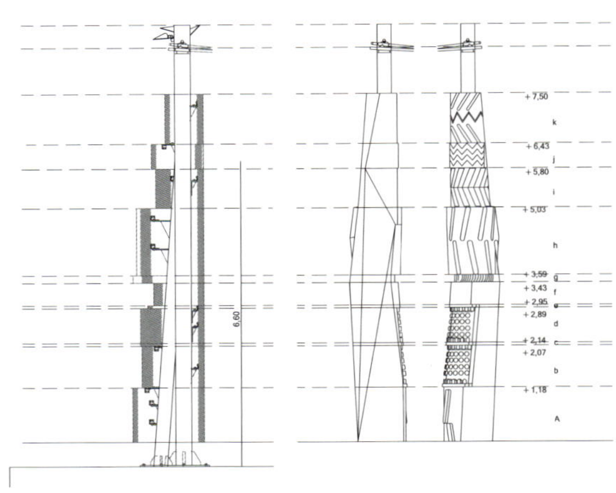

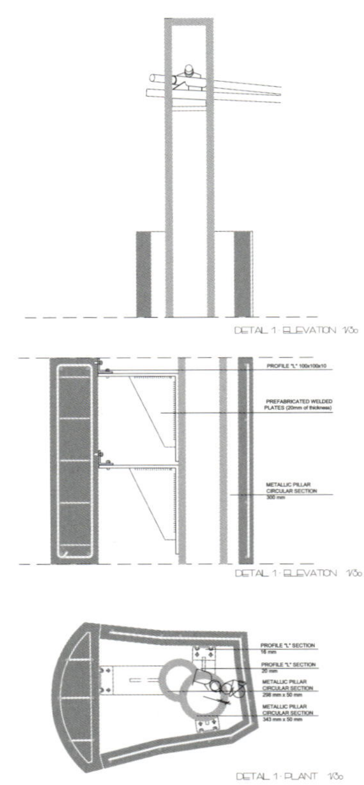

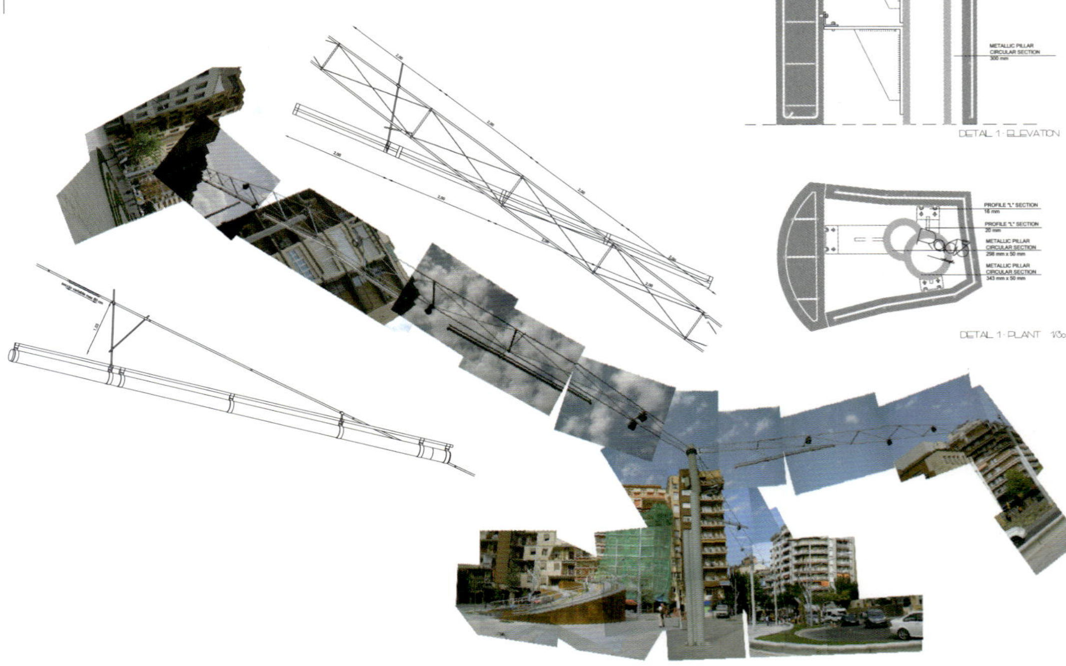

una
millor
CiU

DARLING QUARTER

达令港广场

Sydney, Australia

Darling Harbour South – or "Darling Quarter" as it is now known – is one of Australia's most visited destinations, and the redevelopment of its public domain has been a major place-making transformation project for Sydney.

The 15,000 square meters project includes a retail terrace, a public park, two 6-star commercial buildings, a new pedestrian street, and its centrepiece – at 4,000 m^2 – is the largest children's play space in the Sydney CBD. The unique water play area brings Darling Harbour's industrial heritage to the fore and establishes Darling Quarter as a regional destination.

Two bold pedestrian north-south and east-west links connect the precinct to Sydney city, Chinatown and Cockle Bay. Activated by cafes and retail activities, they become the framework for the public domain design.

• **Water Feature** - Waterforms International

• **Material** - Custom seats within playground – perforated metal with painted (white) finish; Custom seats on edge of playground (pedestrian boulevard) – precast concrete base with timber top; Square bench seats – timber; Bubblers – metal with painted with (micaceous grey) finish; Bins – custom perforated metal with painted (white) finish;

• **Builder** - Bovis Lend Lease

• **Area** -15,000 m^2

• **Photographer** - Florian Groehn, John Marmaras

Darling Quarter

Fleece Street WOOL TAKES OFF
BRIDGE TO THE WEST
GOLD HERALDS A NEW AGE

悉尼达令港南部，也就是今天被人熟知的“达令广场”，是澳大利亚最引人入胜的旅游目的地之一，亦是悉尼极具影响力、最能展现城市新风貌的重点工程项目。

该项目占地 15000 平方米，包括一片零售商业区、一个公园、两座六星级的商业楼宇和一条新建步行街。项目中心是一座 4000 平方米的儿童游乐场，也是悉尼市中心最大的儿童游乐场。独特的水上游戏空间突出了达令港工业港口的特色，将达令港广场变成了澳大利亚最吸引人的旅游胜地。

达令港成功地打造了南北向、东西向两条人行走廊，连接起市中心、唐人街和库克湾。沿街的咖啡馆和零售商铺为街道注入了活力，使其成为了公共空间规划设计的主要框架。

Darling Quarter
UNION SOLUTIONS

OPEN CENTER OF CIVIC ACTIVITIES

Cordova, Spain

市民文化活动中心

A place with the capacity to reconsider the changing needs, aware of its flexible and temporary condition.

The design of the plant cover is closely connected to other points of view from which the action is perceived, both from the existing buildings on the perimeter and from other positions as it happens from the bridge over the tracks of the AVE. Therefore it is of great importance to the treatment of surface color on top.

Equally important has been the ground plane treatment as an activator of different uses. The treated surface looks like a huge game board that can hold any kind of activities, those already known and others yet to be invented. The temporary use of market two days a week is a condition but not in any way a limit, is only one more of the potential uses to be made.

The cover is solved with parasols with diameters between 7 and 15 meters. In turn, the heights are also variable between 4 and 7 meters. This allows great flexibility with little variability and constructive elements. Furthermore, the position at different heights avoids excessive opacity of construction, and allows the passage of light reflection through the various umbrellcs.

The structure is steel umbrellas. Inside the columns is constructed element evacuation of rainwater from the roof, fixed with concrete filling. It is built with aluminum plates in different colors and interior insulation. The treatment of the bottom surface is reflective, giving the place a huge potential in the way that light activates the color by reflection.

Finally note that artificial lighting is fixed in the same parts, relying on a uniform light on the ground plane and the capture of light by reflection from the undersides of the parasols, light points are in different orbits each parasol taking into account the amount of light that reaches the sun shades nearby.

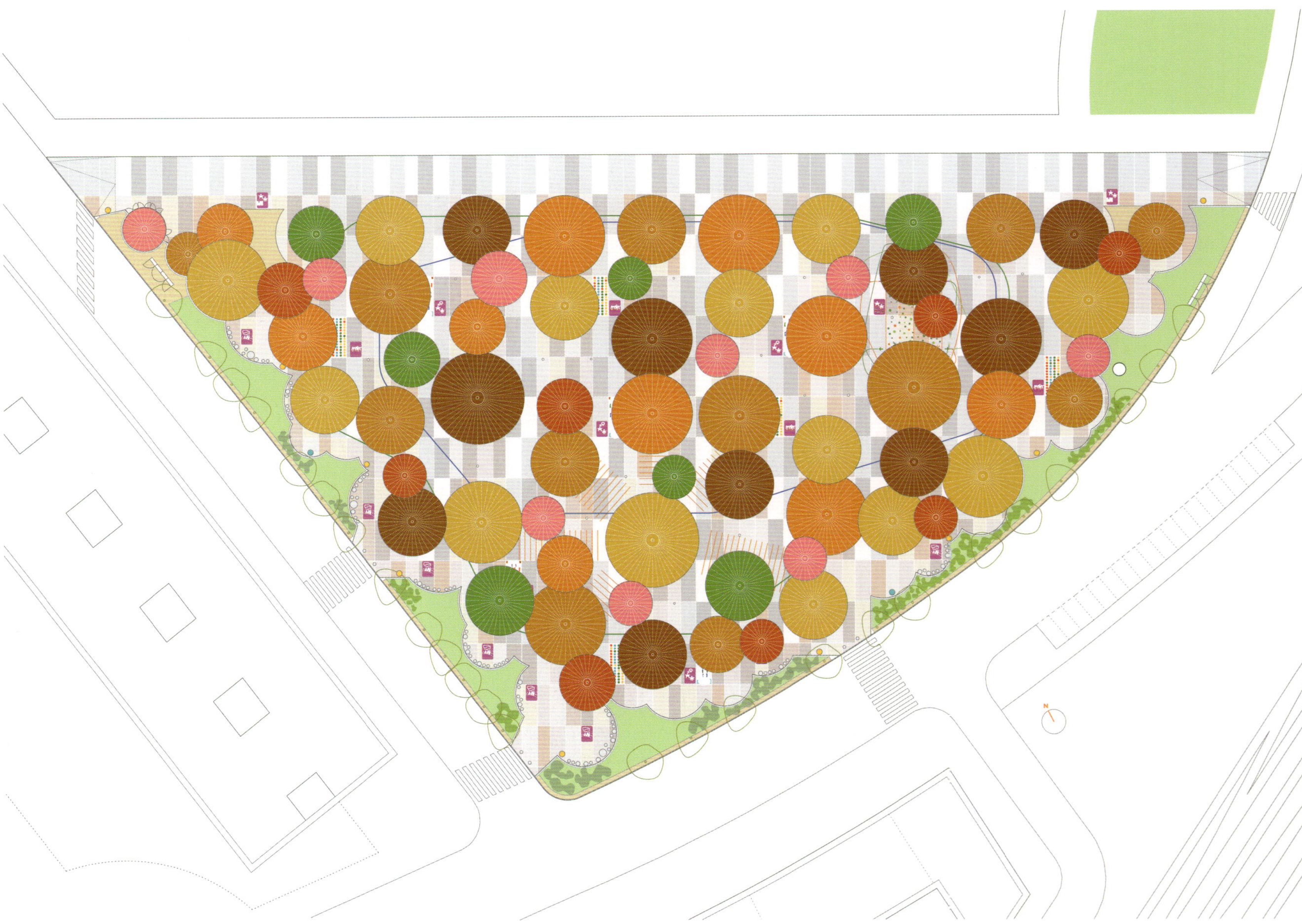

• **Design Team** - Fernando G.Pino, Manuel G. de Paredes, Raquel Blasco, David Pérez • **Area** - 11,920 m² • **Photographer** - Paredes Pino

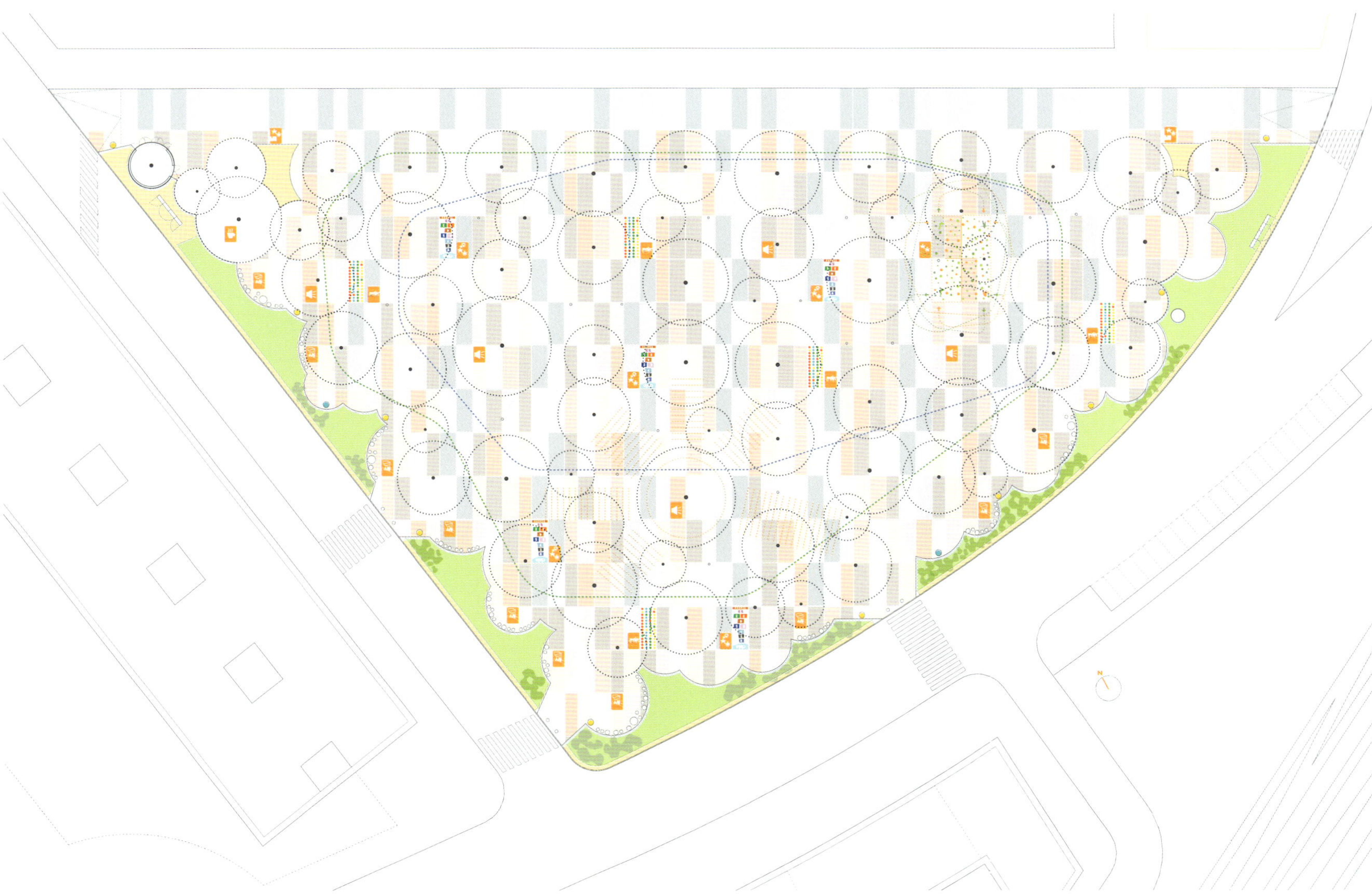

PISA EL COLOR

C/ ISLAS SISARGAS
C/ ISLA DE FORMENTERA
C/ ISLAS CIES
100.40
100.20
100.00
99.80
99.70
100.30
100.00
99.80
99.70
N
CHILDREN PLAYGROUND AREAS
LAGOON GAME. 1 UNIT
RAYUELA GAME. 5 UNITS
ENREDOS GAME. 5 UNITS
CAFETERIA
LOUDSPEAKER SYSTEM
SPORT AREAS
FOOTING TRACK. 250 METERS
FOOTING TRACK 300 METERS
SEATING AREAS

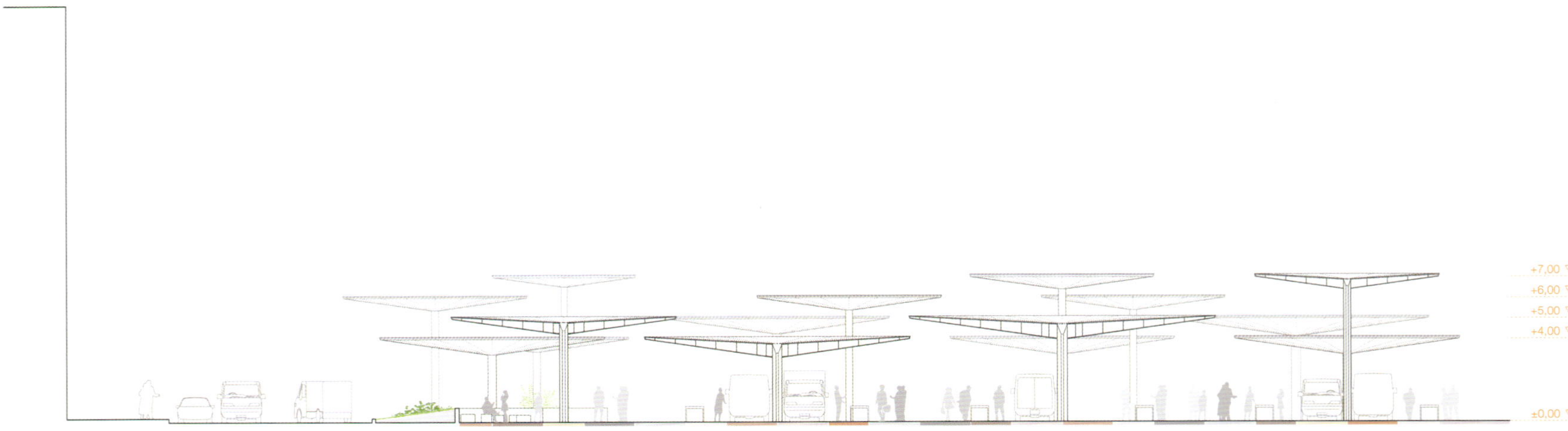
+7,00
+6,00
+5,00
+4,00
±0,00

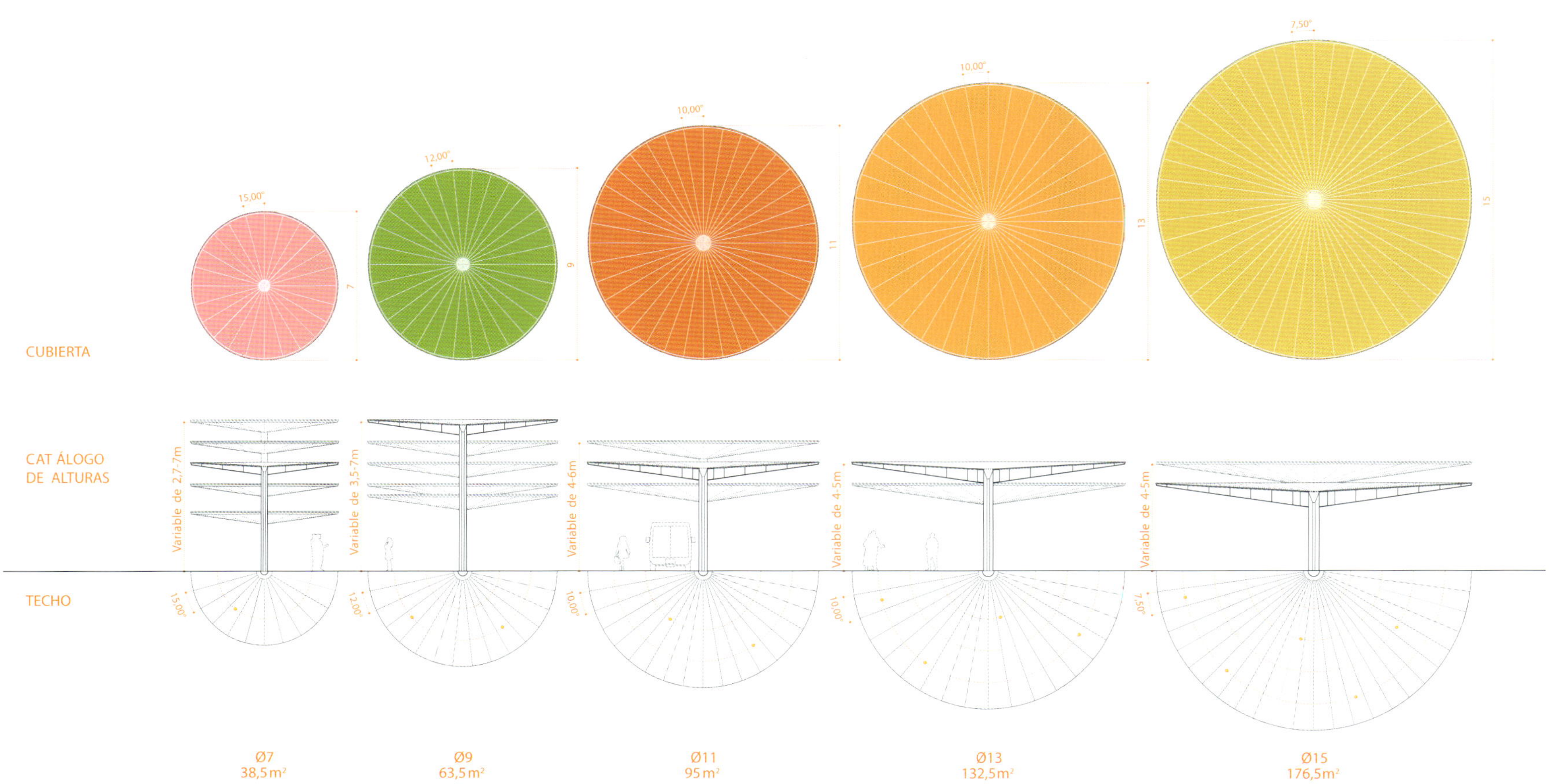

CUBIERTA
15,00°
12,00°
10,00°
10,00°
7,50°
CAT ÁLOGO DE ALTURAS
Variable de 2,7-7m
Variable de 3,5-7m
Variable de 4-6m
Variable de 4-5m
Variable de 4-5m
TECHO
Ø7 38,5m²
Ø9 63,5m²
Ø11 95m²
Ø13 132,5m²
Ø15 176,5m²

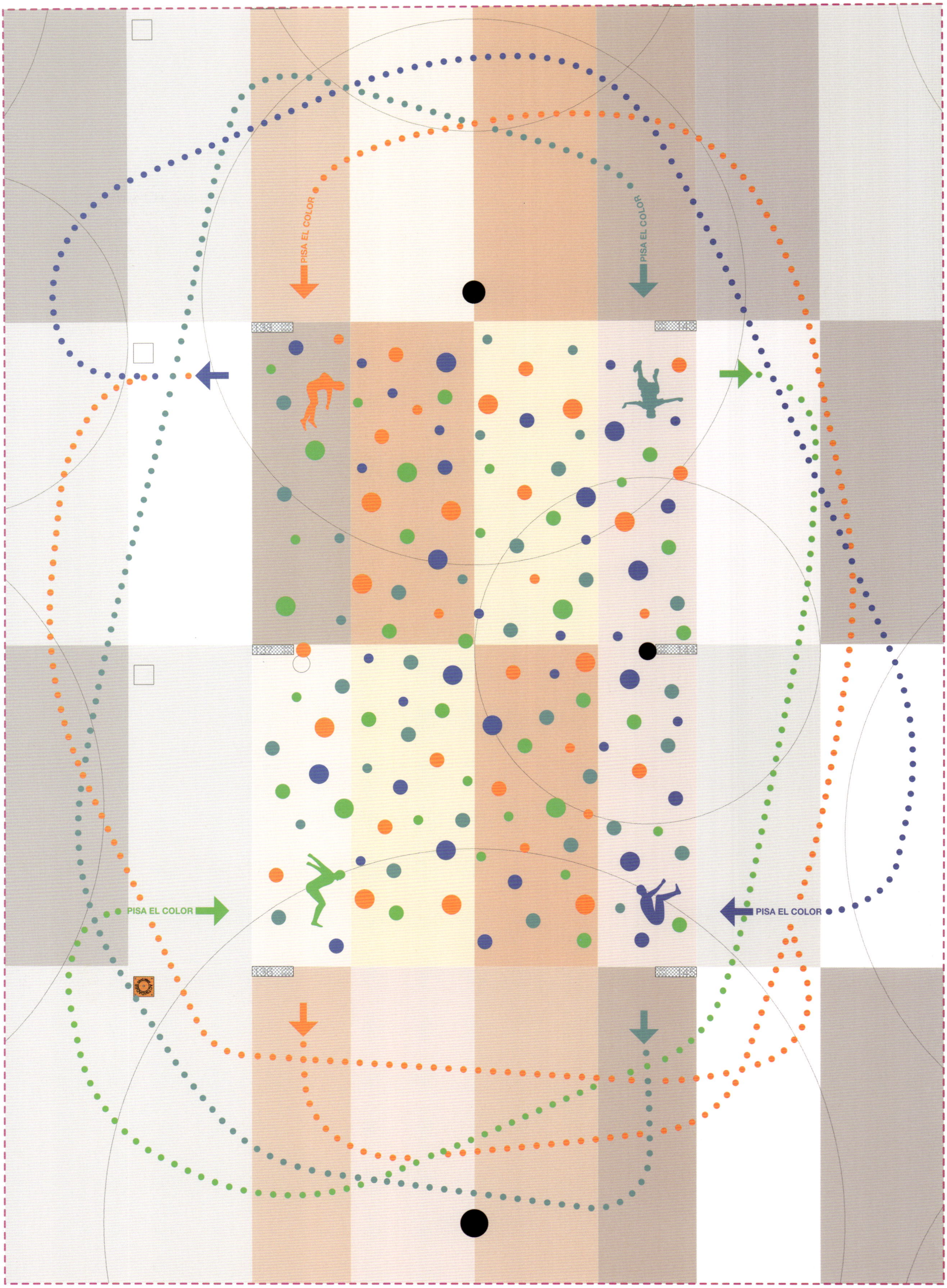
PISA EL COLOR
PISA EL COLOR
PISA EL COLOR
PISA EL COLOR

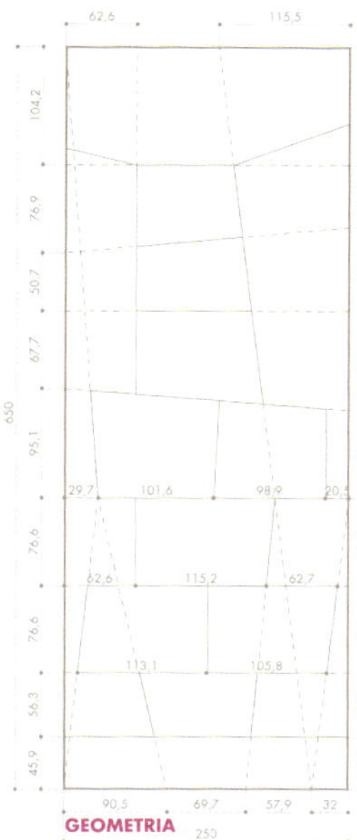

GEOMETRIA

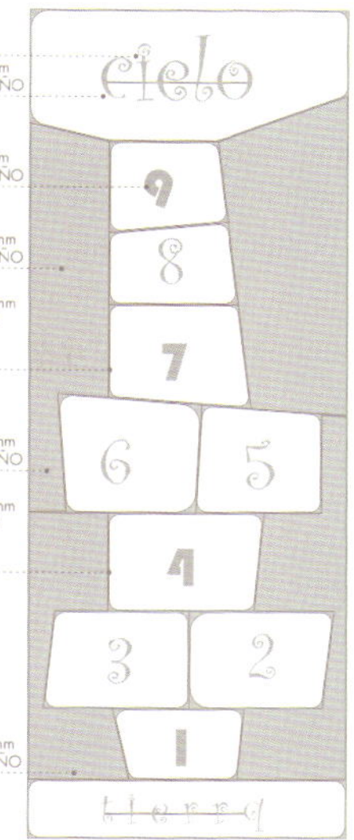

MARCOS DE ACERO INOXIDABLE

RELLENO ENTRE MARCOS DE ACERO INOXIDABLE A BASE DE PAVIMENTO CONTINUO DE HORMIGÓN COLOREADO

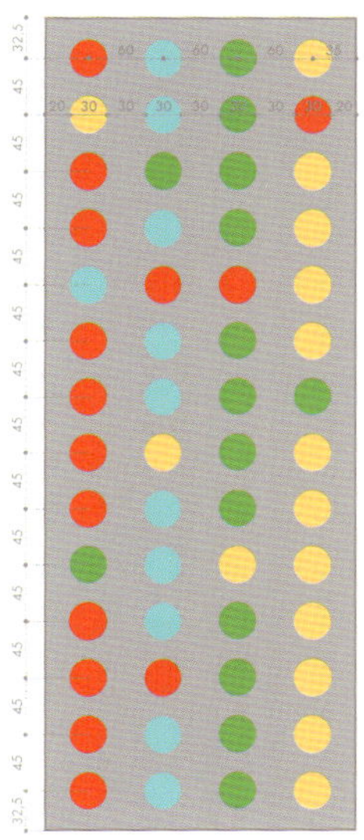

GEOMETRIA

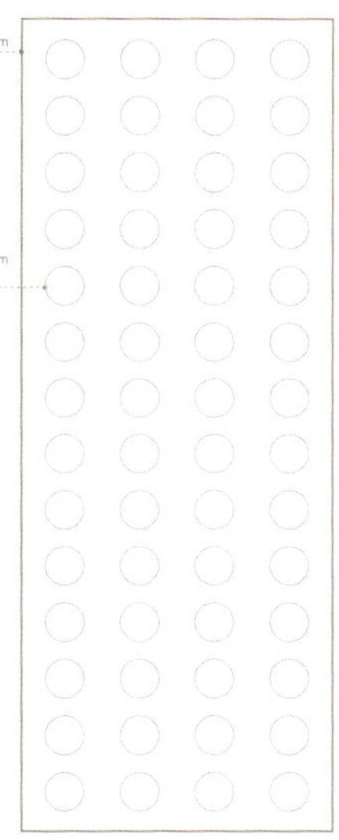

PIEZAS DE ACERO INOXIDABLE

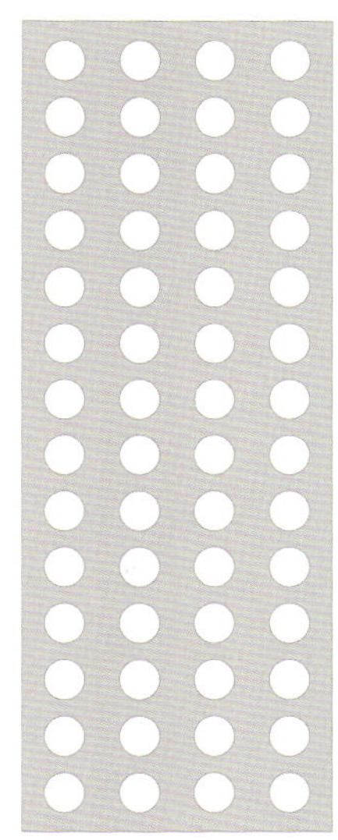

RELLENO ENTRE MARCOS CIRCULARES DE ACERO INOXIDABLE A BASE DE PAVIMENTO CONTINUO DE HORMIGÓN COLOREADO EN NEGRO

PIEZAS CIRCULARES DE PIEDRA EN COLOR A DEFINIR POR LA DF

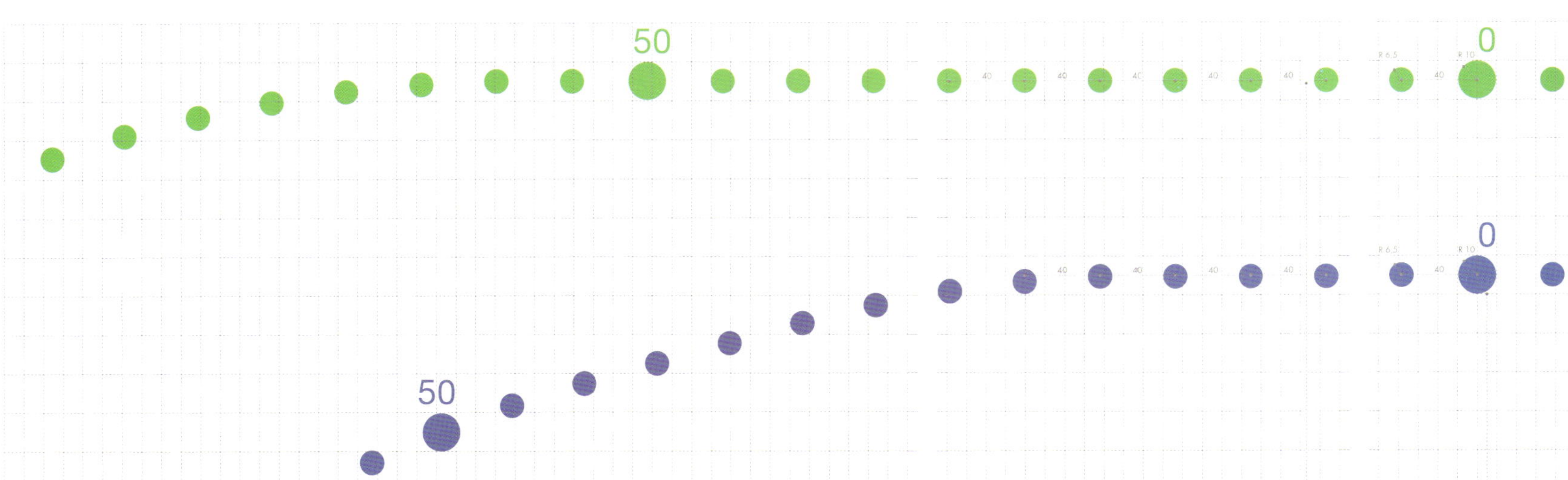

TODA LA PINTURA SEÑALIZACION HORIZONTAL SERA DE ALTA RESISTENCIA EN ACABADO MATE

- COLOR NARANJA RAL 2003
- COLOR AZUL RAL 5015
- COLOR VERDE RAL 6032
- COLOR VERDE RAL 6033

RECORRIDO DE 300 METROS DE LONGITUD MARCADO MEDIANTE SUCESIÓN CÍRCULOS DE Ø13CM CADA 40 CM. REALIZADOS PINTURA DE SEÑALIZACIÓN HORIZONTAL DE ALTA RESISTENCIA ACABADO MATE COLOR VERDE RAL 6032. INDICACIÓN DE LONGITUD DE RECORRIDO CADA 50 M MEDIANTE CIRCULOS DE Ø20CM EN EL MISMO COLOR, TEXTO TIPOLOGÍA HELVÉTICA NEUE, CUERPO DE LETRA 13CM

RECORRIDO DE 300 METROS DE LONGITUD MARCADO MEDIANTE SUCESIÓN CÍRCULOS DE Ø13CM CADA 40 CM. REALIZADOS PINTURA DE SEÑALIZACIÓN HORIZONTAL DE ALTA RESISTENCIA ACABADO MATE COLOR AZUL RAL 5015. INDICACIÓN DE LONGITUD DE RECORRIDO CADA 50 M MEDIANTE CIRCULOS DE Ø20CM EN EL MISMO COLOR, TEXTO TIPOLOGÍA HELVÉTICA NEUE, CUERPO DE LETRA 13CM

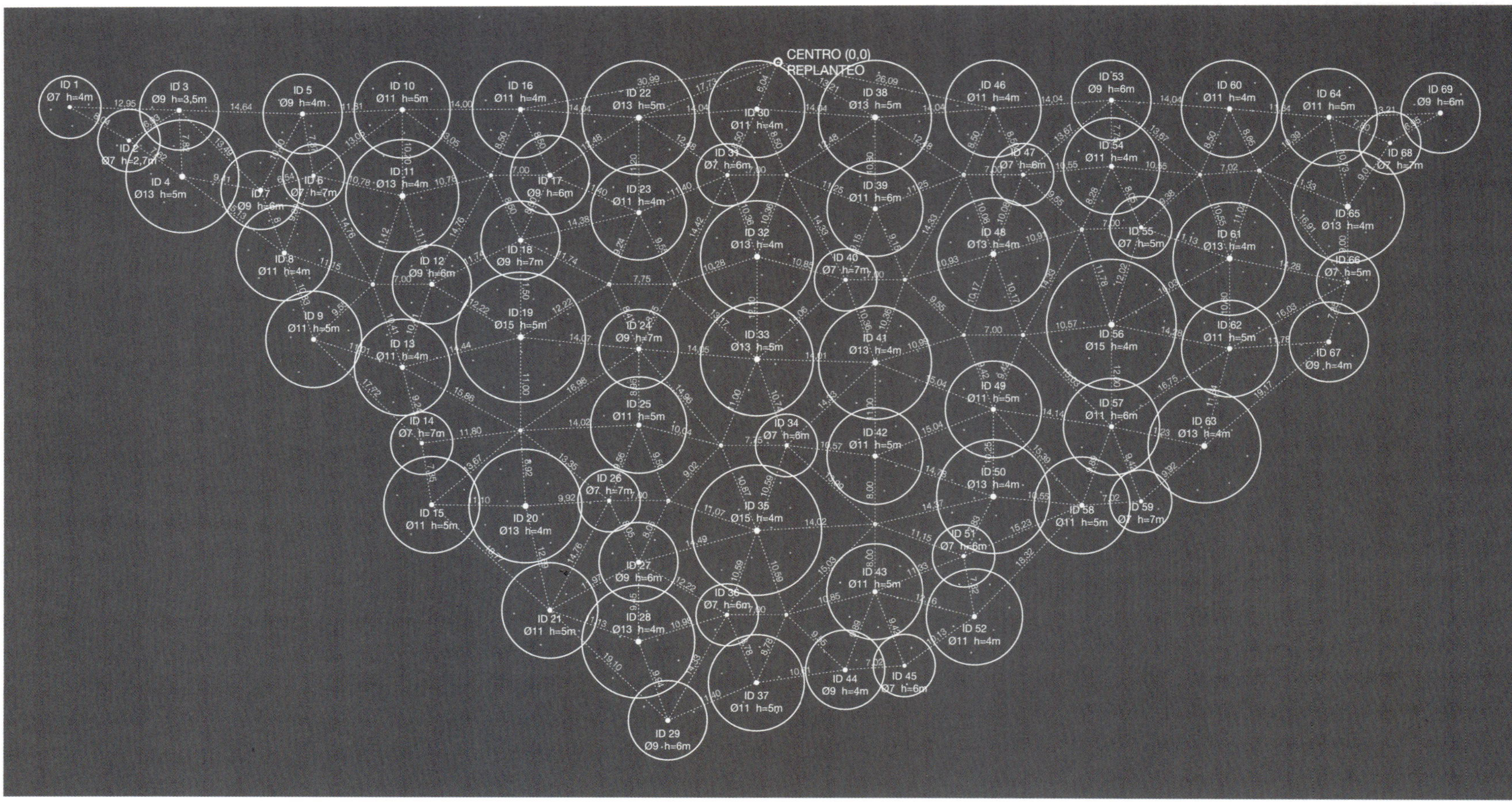

CAAC USES AND CONFIGURATIONS

MOBILE MARKET SET
168 street vendor kiosk, 168 vehicles

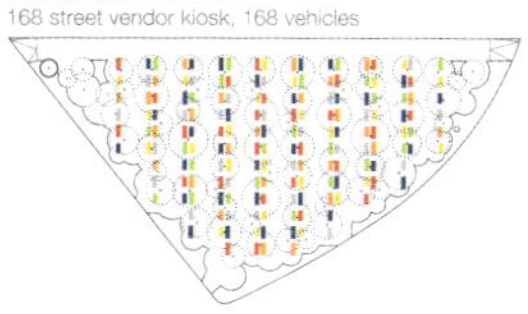

SETTING WITH STAGE

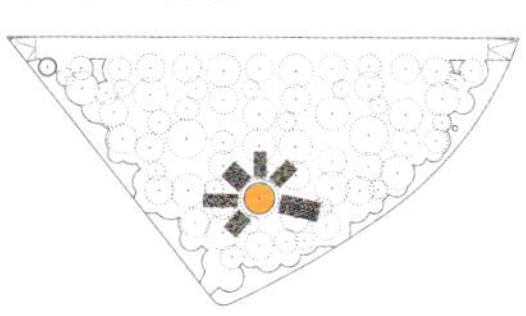

FASHION PARADE SETTING

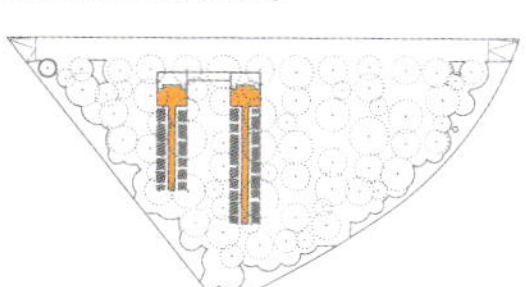

GASTRONOMIC MEETING

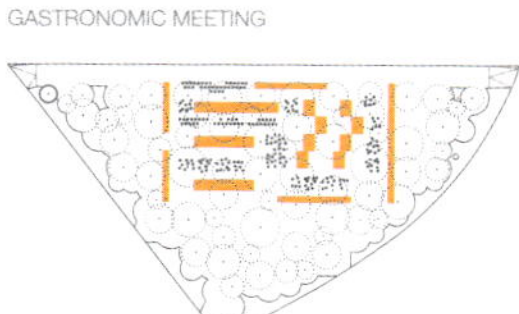

BICYCLE AND KART RACING

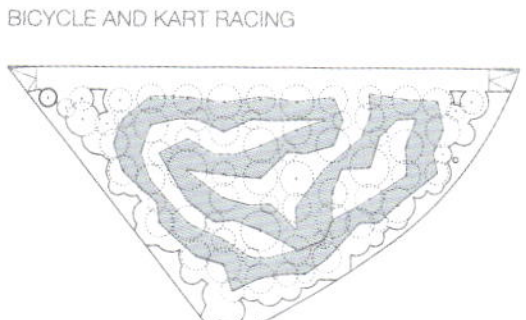

SUMMER OUTDOR CINEMA SETTING

CAAC MOBILITY

MARKET DAY MOBILITY

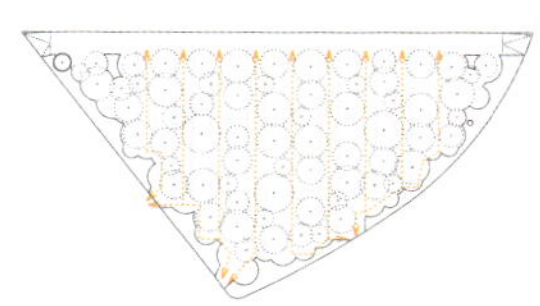

REST OF THE WEEK

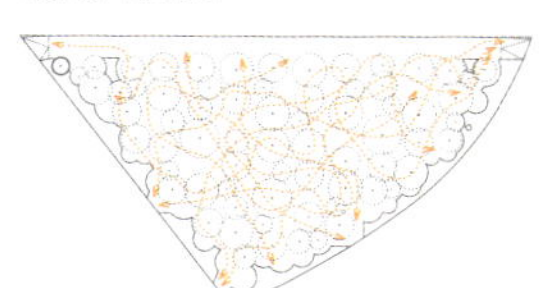

VEHICULAR TRAFFIC

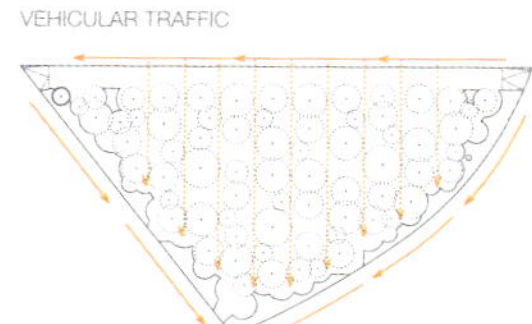

CAAC COMPONENTS OF THE SYSTEM

PETALS UPPER DECK

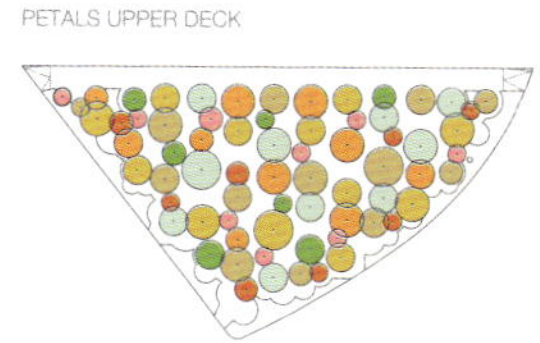

ACTIVABLE PAVEMENT

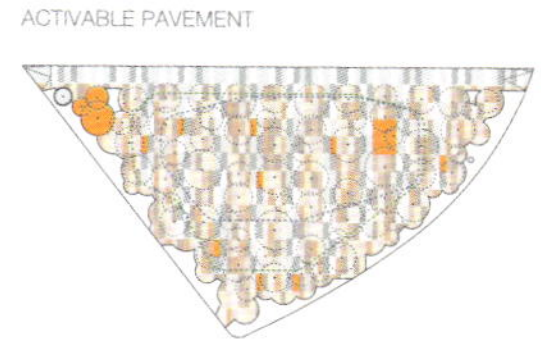

GREEN AREAS

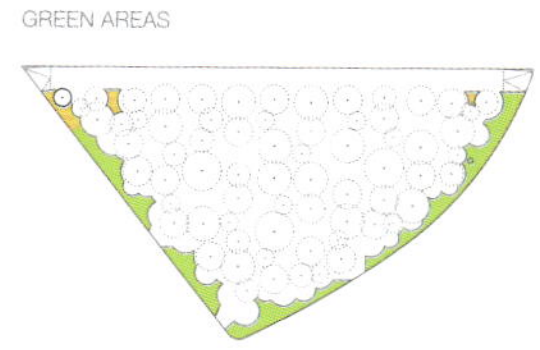

CAAC PREFABRICATION

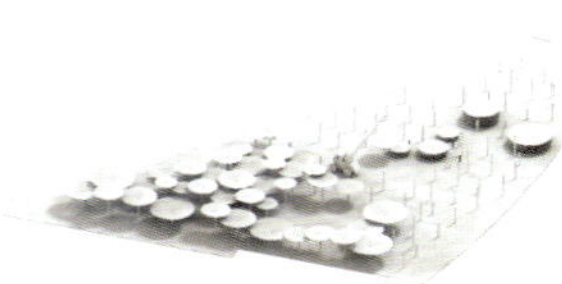

西班牙科尔多瓦市的市民文化活动中心在设计规划上充分考虑了城市公众不断增加的需求，特别强调通过临时改变内部设施，灵活有效地利用空间。

设计师们使用绿色植物覆盖了活动中心建筑，并与周边其他建筑紧密相连，这样不论是从原有的建筑还是其他角度看去，都像是一座天桥横跨在大街上方。因此，建筑表面的设计处理十分重要。

同样重要的还包括基于不同用途对地面进行的平整工程。经过平整后的地面如同一个巨大的游戏平台，可以在上面进行各种活动。这里还可作为临时商品交易区，每周两天，而这仅仅是亟待开发的空间用途之一。

地面由直径为 7 至 15 米，高度为 4 至 7 米的多把太阳伞覆盖。这样的处理方法使空间的使用更加灵活，同时不需要进行太多设施上的改造工程。另外，不同的高度设计减少了建筑本身的不透明部分，使灯光在许多太阳伞之间反射。

建筑结构主要采用钢制伞状构造，内部有引导雨水从顶部流下的结构，使用混凝土进行固定。建筑外部由不同颜色具有绝缘内层的铝板作为装饰。底部的表面主要是反光材料，利用光线的反射营造宏大的视觉效果。

最后，需要指出的是均匀排布的人工照明系统，地面上规格相同的灯光，加上地面太阳伞的反射，形成的光斑照射在不同大小的太阳伞下的阴暗处，显得五彩缤纷，令人眼花缭乱。

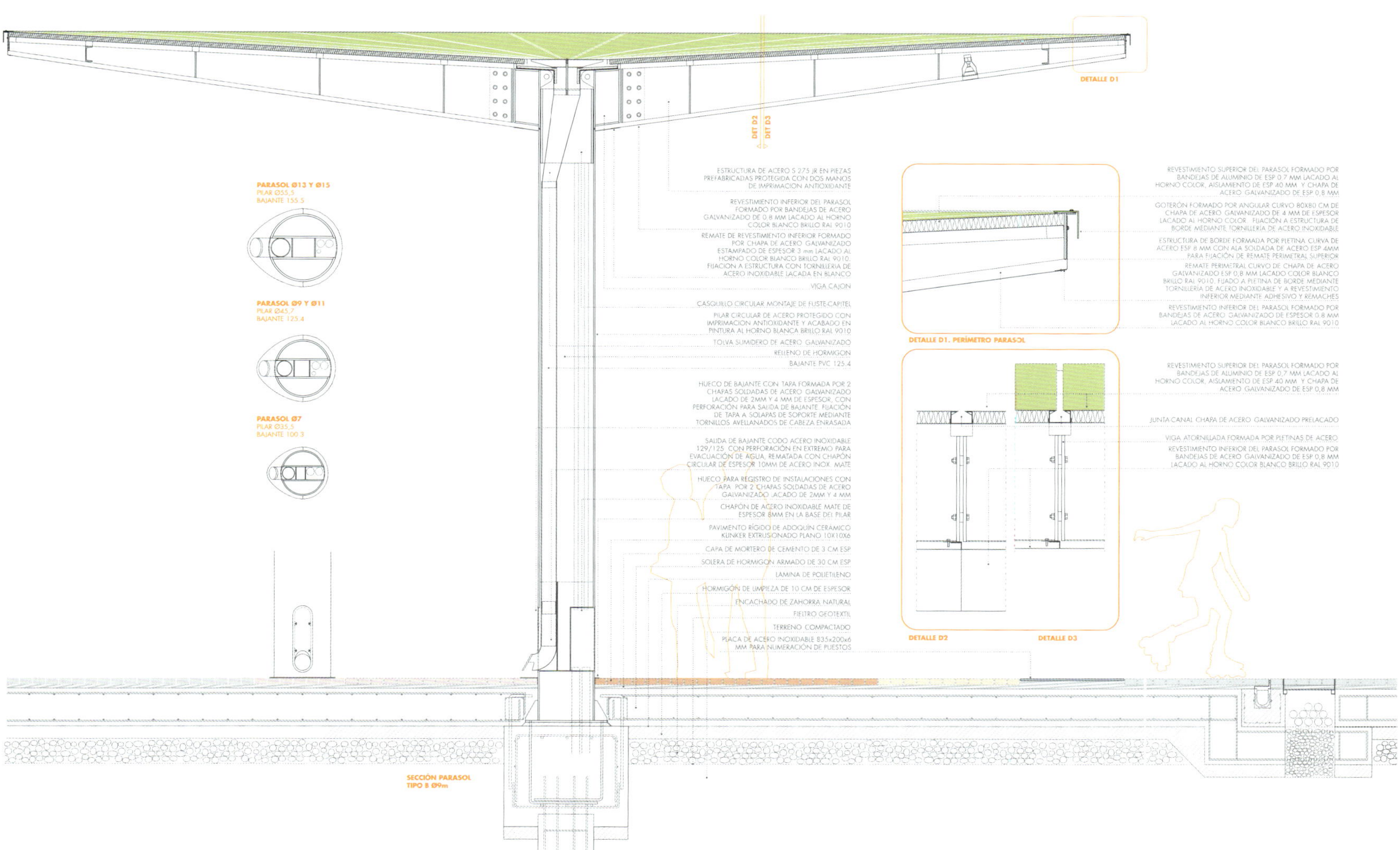
DETALLE D1
DET D2
DET D3
PARASOL Ø13 Y Ø15
PILAR Ø55,5
BAJANTE 155.5
PARASOL Ø9 Y Ø11
PILAR Ø45,7
BAJANTE 125.4
PARASOL Ø7
PILAR Ø35,5
BAJANTE 100.3
ESTRUCTURA DE ACERO S 275 JR EN PIEZAS PREFABRICADAS PROTEGIDA CON DOS MANOS DE IMPRIMACION ANTIOXIDANTE
REVESTIMIENTO INFERIOR DEL PARASOL FORMADO POR BANDEJAS DE ACERO GALVANIZADO DE 0,8 MM LACADO AL HORNO COLOR BLANCO BRILLO RAL 9010
REMATE DE REVESTIMIENTO INFERIOR FORMADO POR CHAPA DE ACERO GALVANIZADO ESTAMPADO DE ESPESOR 3 mm LACADO AL HORNO COLOR BLANCO BRILLO RAL 9010. FIJACION A ESTRUCTURA CON TORNILLERIA DE ACERO INOXIDABLE LACADA EN BLANCO
VIGA CAJON
CASQUILLO CIRCULAR MONTAJE DE FUSTE-CAPITEL
PILAR CIRCULAR DE ACERO PROTEGIDO CON IMPRIMACION ANTIOXIDANTE Y ACABADO EN PINTURA AL HORNO BLANCA BRILLO RAL 9010
TOLVA SUMIDERO DE ACERO GALVANIZADO
RELLENO DE HORMIGON
BAJANTE PVC 125.4
HUECO DE BAJANTE CON TAPA FORMADA POR 2 CHAPAS SOLDADAS DE ACERO GALVANIZADO LACADO DE 2MM Y 4 MM DE ESPESOR, CON PERFORACIÓN PARA SALIDA DE BAJANTE. FIJACIÓN DE TAPA A SOLAPAS DE SOPORTE MEDIANTE TORNILLOS AVELLANADOS DE CABEZA ENRASADA
SALIDA DE BAJANTE CODO ACERO INOXIDABLE 129/125. CON PERFORACIÓN EN EXTREMO PARA EVACUACIÓN DE AGUA, REMATADA CON CHAPÓN CIRCULAR DE ESPESOR 10MM DE ACERO INOX. MATE
HUECO PARA REGISTRO DE INSTALACIONES CON TAPA POR 2 CHAPAS SOLDADAS DE ACERO GALVANIZADO LACADO DE 2MM Y 4 MM
CHAPÓN DE ACERO INOXIDABLE MATE DE ESPESOR 8MM EN LA BASE DEL PILAR
PAVIMENTO RÍGIDO DE ADOQUÍN CERÁMICO KLINKER EXTRUSIONADO PLANO 10X10X6
CAPA DE MORTERO DE CEMENTO DE 3 CM ESP
SOLERA DE HORMIGON ARMADO DE 30 CM ESP
LAMINA DE POLIETILENO
HORMIGÓN DE LIMPIEZA DE 10 CM DE ESPESOR
ENCACHADO DE ZAHORRA NATURAL
FIELTRO GEOTEXTIL
TERRENO COMPACTADO
PLACA DE ACERO INOXIDABLE 835x200x6 MM PARA NUMERACIÓN DE PUESTOS
DETALLE D1. PERÍMETRO PARASOL
REVESTIMIENTO SUPERIOR DEL PARASOL FORMADO POR BANDEJAS DE ALUMINIO DE ESP 0,7 MM LACADO AL HORNO COLOR, AISLAMIENTO DE ESP 40 MM Y CHAPA DE ACERO GALVANIZADO DE ESP 0,8 MM
GOTERÓN FORMADO POR ANGULAR CURVO 80X80 CM DE CHAPA DE ACERO GALVANIZADO DE 4 MM DE ESPESOR LACADO AL HORNO COLOR. FIJACIÓN A ESTRUCTURA DE BORDE MEDIANTE TORNILLERÍA DE ACERO INOXIDABLE
ESTRUCTURA DE BORDE FORMADA POR PLETINA CURVA DE ACERO ESP 8 MM CON ALA SOLDADA DE ACERO ESP 4MM PARA FIJACIÓN DE REMATE PERIMETRAL SUPERIOR
REMATE PERIMETRAL CURVO DE CHAPA DE ACERO GALVANIZADO ESP 0,8 MM LACADO COLOR BLANCO BRILLO RAL 9010. FIJADO A PLETINA DE BORDE MEDIANTE TORNILLERÍA DE ACERO INOXIDABLE Y A REVESTIMIENTO INFERIOR MEDIANTE ADHESIVO Y REMACHES
REVESTIMIENTO INFERIOR DEL PARASOL FORMADO POR BANDEJAS DE ACERO GALVANIZADO DE ESPESOR 0,8 MM LACADO AL HORNO COLOR BLANCO BRILLO RAL 9010
REVESTIMIENTO SUPERIOR DEL PARASOL FORMADO POR BANDEJAS DE ALUMINIO DE ESP 0,7 MM LACADO AL HORNO COLOR, AISLAMIENTO DE ESP 40 MM Y CHAPA DE ACERO GALVANIZADO DE ESP 0,8 MM
JUNTA-CANAL CHAPA DE ACERO GALVANIZADO PRELACADO
VIGA ATORNILLADA FORMADA POR PLETINAS DE ACERO
REVESTIMIENTO INFERIOR DEL PARASOL FORMADO POR BANDEJAS DE ACERO GALVANIZADO DE ESP 0,8 MM LACADO AL HORNO COLOR BLANCO BRILLO RAL 9010
DETALLE D2
DETALLE D3
SECCIÓN PARASOL TIPO B Ø9m

BLAS INFANTE SQUARE

布拉斯 · 因方特广场

Lleida, Spain

The proposal represents and symbolizes the new centrality that Blas Infante Square offers in a contemporary way. The project is developed from the duality of the two scales that are mixed: the scale of the neighborhood, immediate environment, and a larger size, the city and its historic center.

For the use, immediate environment has been created a neighborhood place, a place of rest, leisure, neighborhood relationship, with shade and benches, playgrounds, grass and floral vegetation following the route of the pergola as protection for the summer.

The existence of a newly built underground parking at the base of the square has made the integration of service mechanisms unifying one of the objectives of the project. In this way the elevators, stairways exits, ventilation... are plastically integrated playing with the elements of the towers and the pergola, which have been planned to integrate the lighting. This same lighting allows to create different environments, dynamic spaces, spaces of communication, some with more light, others soft lighting with shades of rest and tranquility. The public space has to pamper the user.

The impressive topography of the Historic Center of Lleida, consisting of steep streets around the hill from the cathedral of La Seu Vella (1204 AD) is the reference nearly ubiquitous. They assume a flat square with a gentle slope that is picking up as it deviates from the riba of the river. From this public space the entire Old City is shown as it really is, with its grandeur and its conflicts. It is a vantage point, an excellent viewpoint of the city, presenting its prettiness and sincereness.

• **Collaborator** - Nagore Linares, Relja Ferusic, Marina Daviu, Charlotte Devaux, Uxía Carballeira

• **Constructor** - Acsa Sorigué

• **Area** - 8,403 m^2

• **Photographer** - Jordi Bernadó, Cristobal Bryjowski

INOUT
INOUT

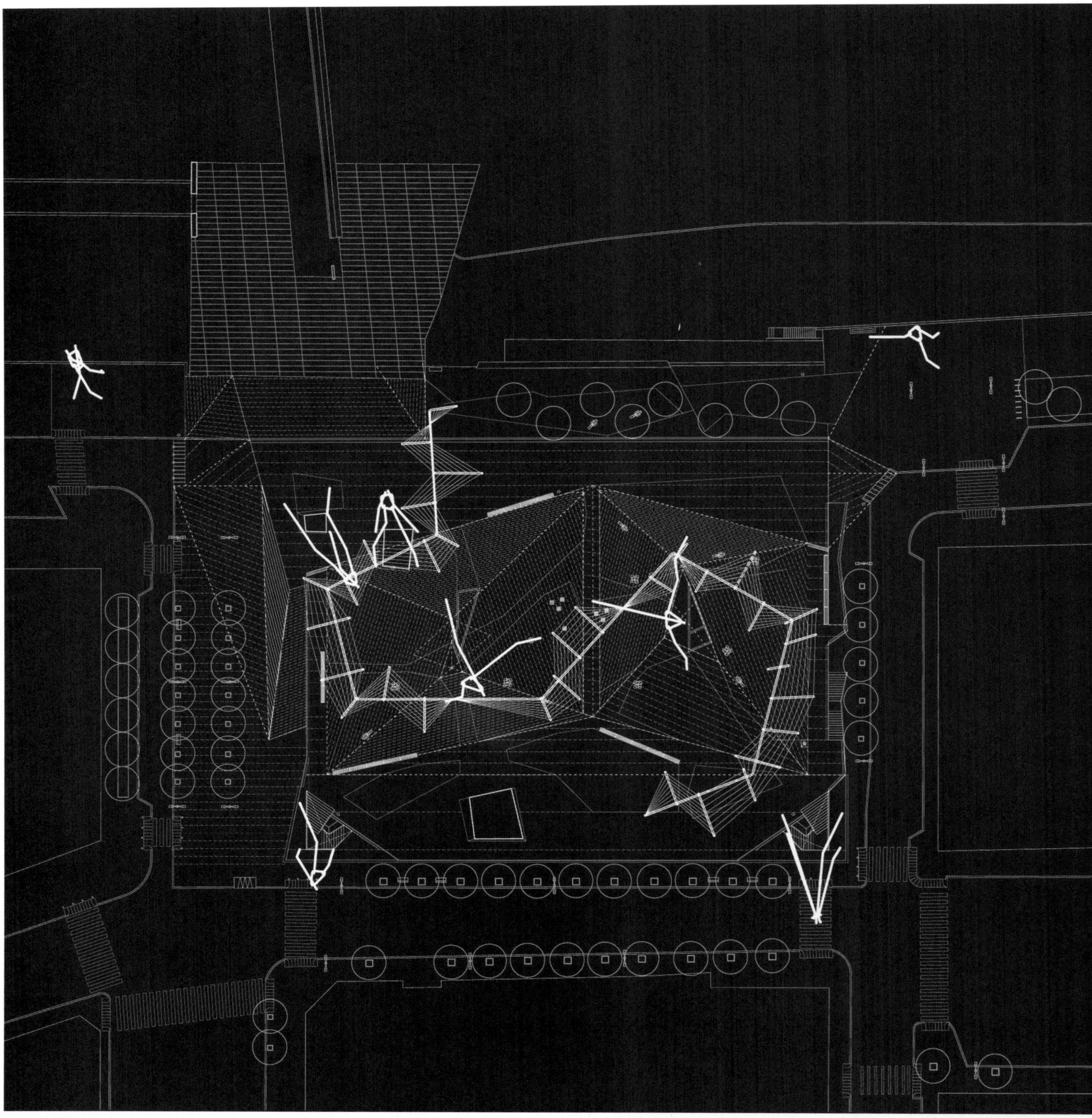

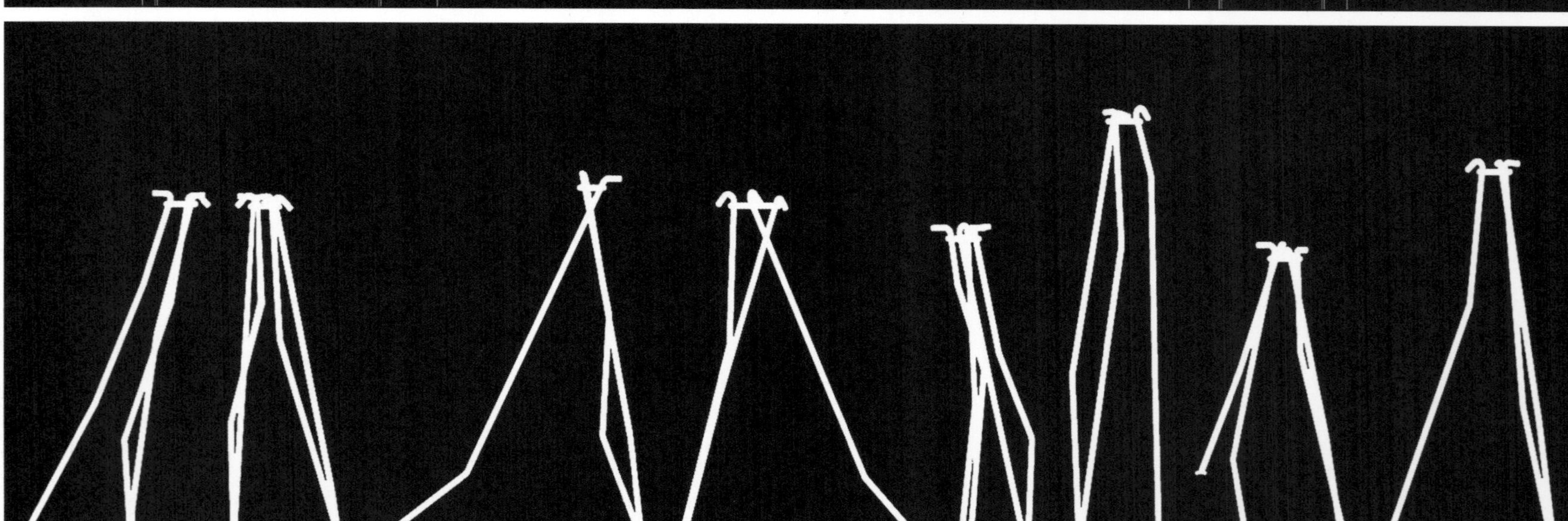

d'aire i de fe, la dura veu de bronze

布拉斯·因方特广场属于典型的当代建筑风格，广场设计方案代表并象征着新的城市核心的诞生。项目的开发设计突出展现两个层次相混合的二元性，一是周边的直接环境，二是作为城市和历史中心的重要地位。

根据项目的需要，周边的直接环境已经被改建为一个亲民的社区，同时也是一个休闲、娱乐、融合邻里关系的场所。这里专门设计了遮阳的树木、休闲用的长凳和游乐场，人行小路穿梭在草地和植被中间，上面设有藤架，夏天藤蔓缠绕，遮蔽阳光。

广场的地下新建了一个停车场，通过这里可以将整个广场项目中的服务项目整合在一起。这样，电梯、楼梯口、通风等功能元素，加上高塔、小路，通过自然光线的照射融合在一起，形成一个有机的整体。利用光线的变换，还能营造出不同的环境、不同的动态空间以及不同的交流空间。有些空间需要强烈的光线，还有些如休闲和安静的地方则适合柔和的光线环境。公共空间可以满足不同使用者的要求。

莱里达历史中心周边的地形令人印象十分深刻，例如，从建于公元1204年的塞乌维拉大教堂引出的环山公路十分陡峭，类似这样的地形比比皆是。这些不同的地形构成了一个带有平缓坡度的平面方形，从利巴河岸向远方逐渐延伸。整个古城的公共空间真实地表现出极度的宏伟壮观，同时也客观地体现了与现代风格的激烈冲突。这里地理位置优越，拥有观赏整座城市的最佳视角，城市的美丽和真诚直接呈现在眼前。

MÅLØV AXIS

马洛夫中轴区

Ballerup, Denmark

Måløv Axis and the adjacent city space is a visionary project underlining the lively character of Måløv city in close contact to the amazing surrounding landscape. The movement trough the axis will – beyond being easy accessible for everyone, safe and clear – provide a spatial story relating to the distinctive character of Måløv and create an experience for citizens and visitors.

The main idea sets off in a reading of the wide landscape and natural context in this area. The landscape is a characteristic moraine landscape shaped through the impact of the glacial period. The glaciers have left a rolling landscape with kept ice holes, valleys and plains. A beautiful and recreational landscape accessible from North via Måløv Nature Park and from South via Søndergaard Lake and the Måløv wedge.

As a result the Måløv Axis will appear as a fluent landscape building on the great historical values in the landscape creating a vibrant area full of city life and natural environments, rich on experiences, interesting and safe for everyone. A new connecting urban space in the city will interwine the existing and the new values.

The Måløv Axis was awarded the Danish Landscape Award 2010 as well as the Ballerup Municipality Building Award 2011.

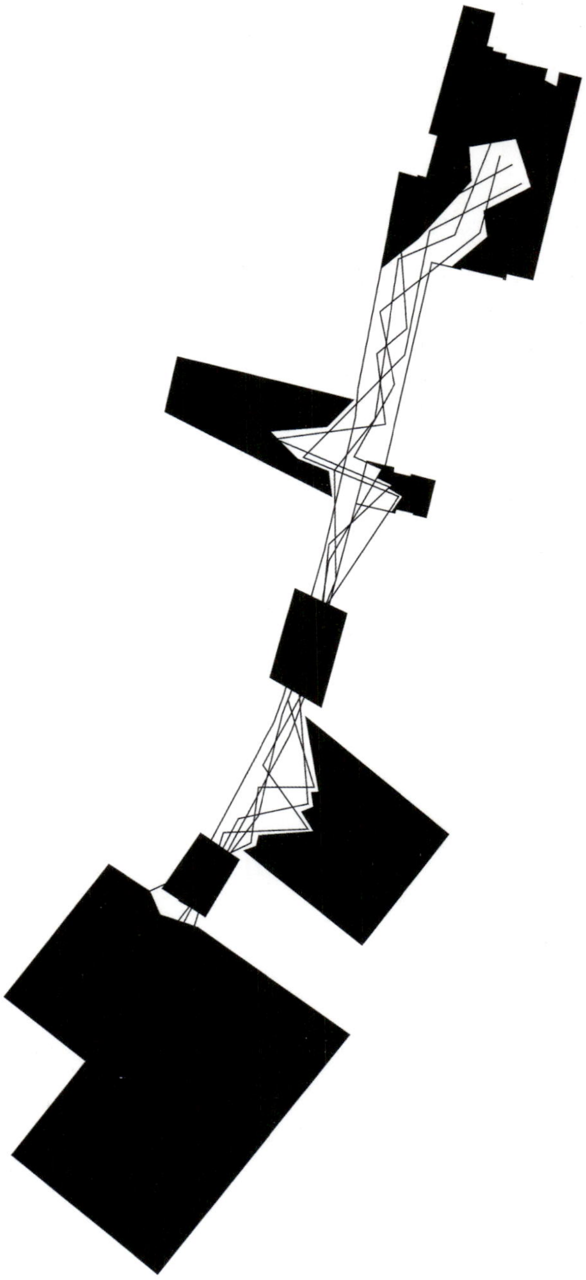

马洛夫中轴区和邻近的城市空间共同构筑了一个风景宜人的区域，结合周围美丽的景观，着重突出了马洛夫市灵动的城市规划风格。中轴区的规划十分便于公众通行，安全洁净，利用一种独特的空间布局，为马洛夫市的市民和游客提供了一种与大自然亲密接触的体验和感受。

设计师们通过分析广场宽敞的地域和周边的自然环境，确定了项目设计方案的主要理念。马洛夫市所在地的地形由冰川时期的流水冲击而成，是典型的冰碛景观。冰川运动留下了一种起伏的地形景观，伴有冰洞、峡谷和平原等。中轴区北邻马洛夫自然公园，南接市郊的桑德加德湖，形成一个集秀丽的风景与休闲娱乐为一体的公共区域。

项目完成后，这里将为公众展现一个极具历史意义的流畅的景观带，打造一种更为休闲的都市生活节奏和自然环境，让每个市民都能感受到安全与舒适。中轴区连接不同的城市空间，将原有的建筑与新的景观融为一体。

马洛夫中轴区项目获得了 2010 年度丹麦景观设计奖和 2011 年度巴勒鲁普自治市建筑大奖。

MÅLØV

MÅLØV

LUX – PLACE DE L'ACADÉMIE

卢克斯广场

Esch-sur-Alzette, Luxembourg

AllesWirdGut are delighted to announce that the project LUX – Place de l'Académie, the architectural design on a former "Stahlhof" (steel industry site) in Esch-sur-Alzette, won the prestigious Luxembourg "Bauhaereprais" (builders award) in the category "Espaces paysages" (space and landscape design). The award was accepted yesterday, 26 June 2012 by the builders of AGORA and AllesWirdGut at a festive presentation in Luxembourg.

The atmosphere of the abandoned industrial site was characterized by width and roughness, and by plants such as mosses and birches. The redesign by AllesWirdGut is very gentle with this legacy, leaving the hidden qualities of the former site shine through to the present. Aging-capable materials such as concrete, wood and untreated steel were combined with an industrial building-like architecture. Space alternates with empty seats, steps and landscaping. The materials used take on patina with age and allow the return of the past.

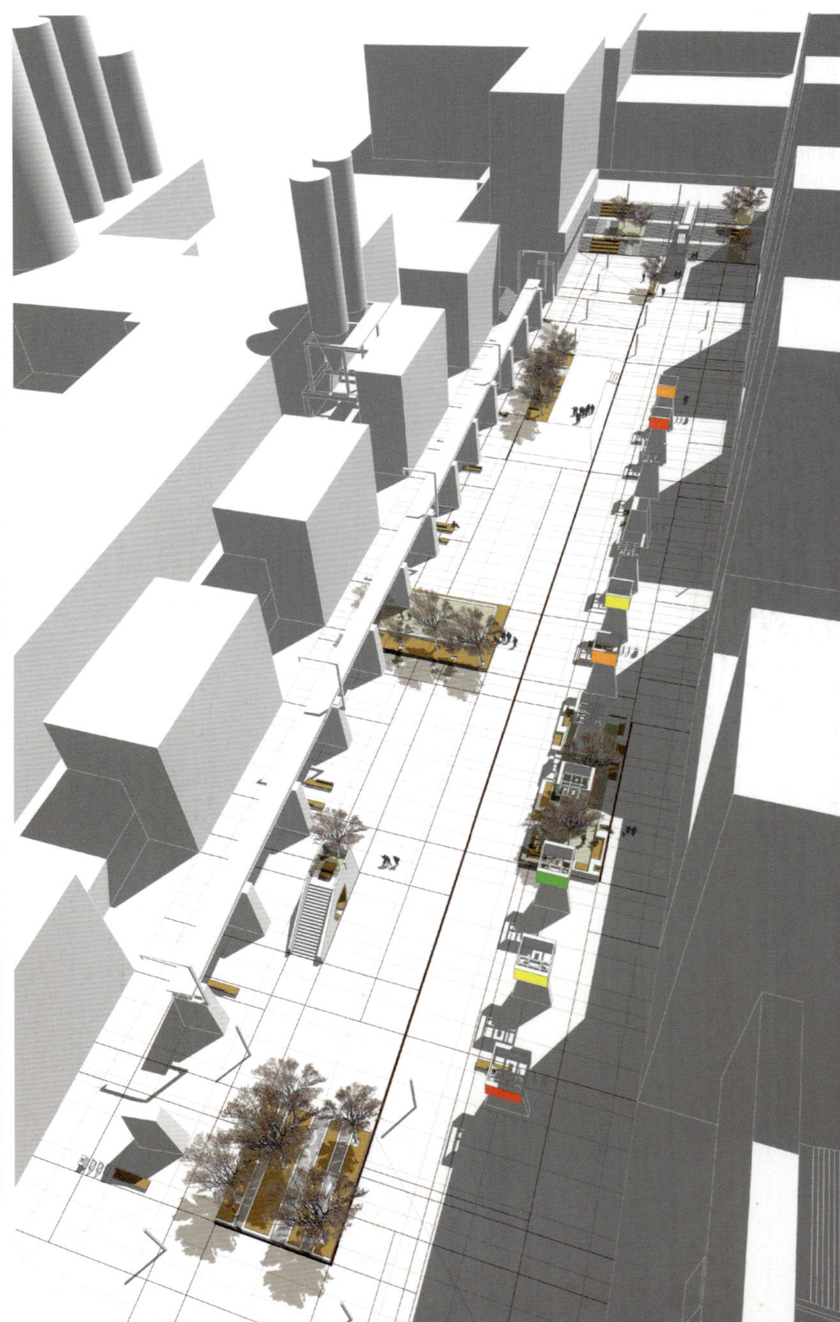

BELVAL
BELVAL

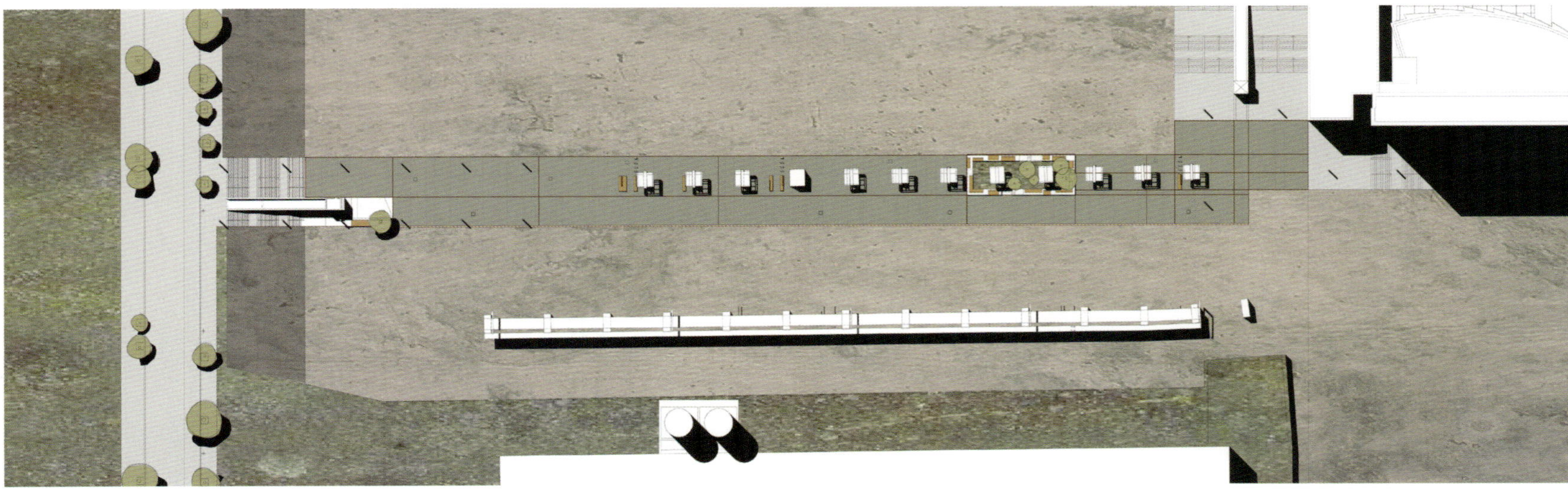

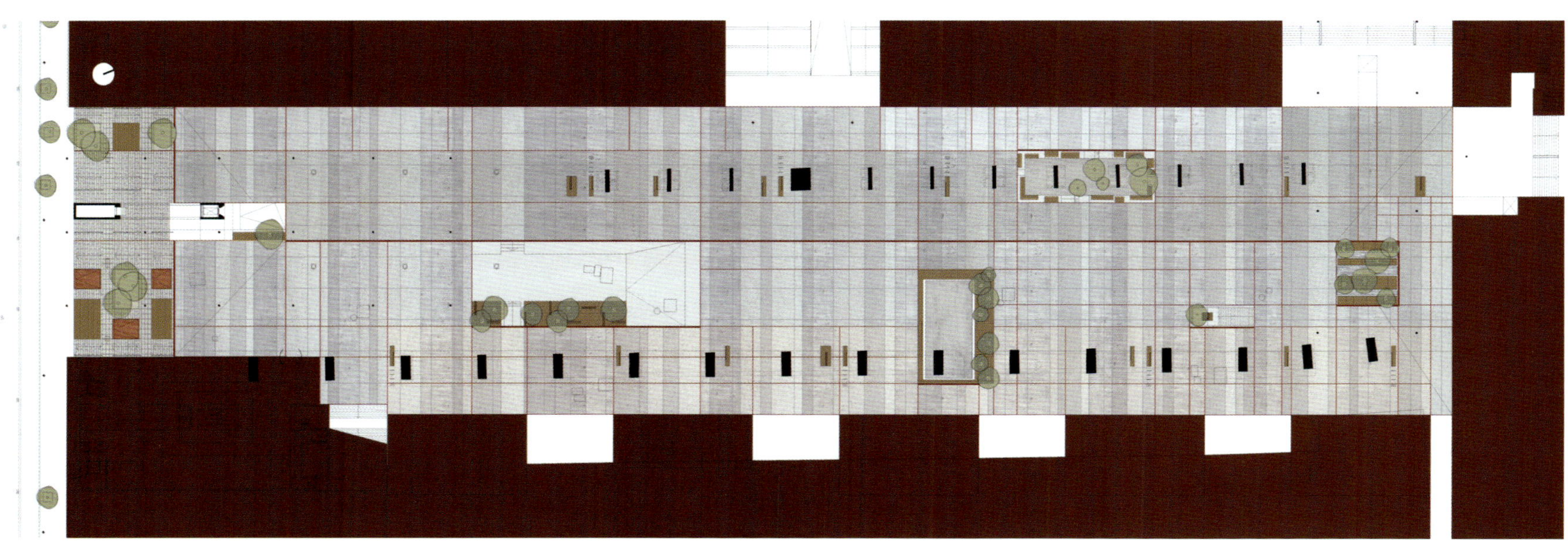

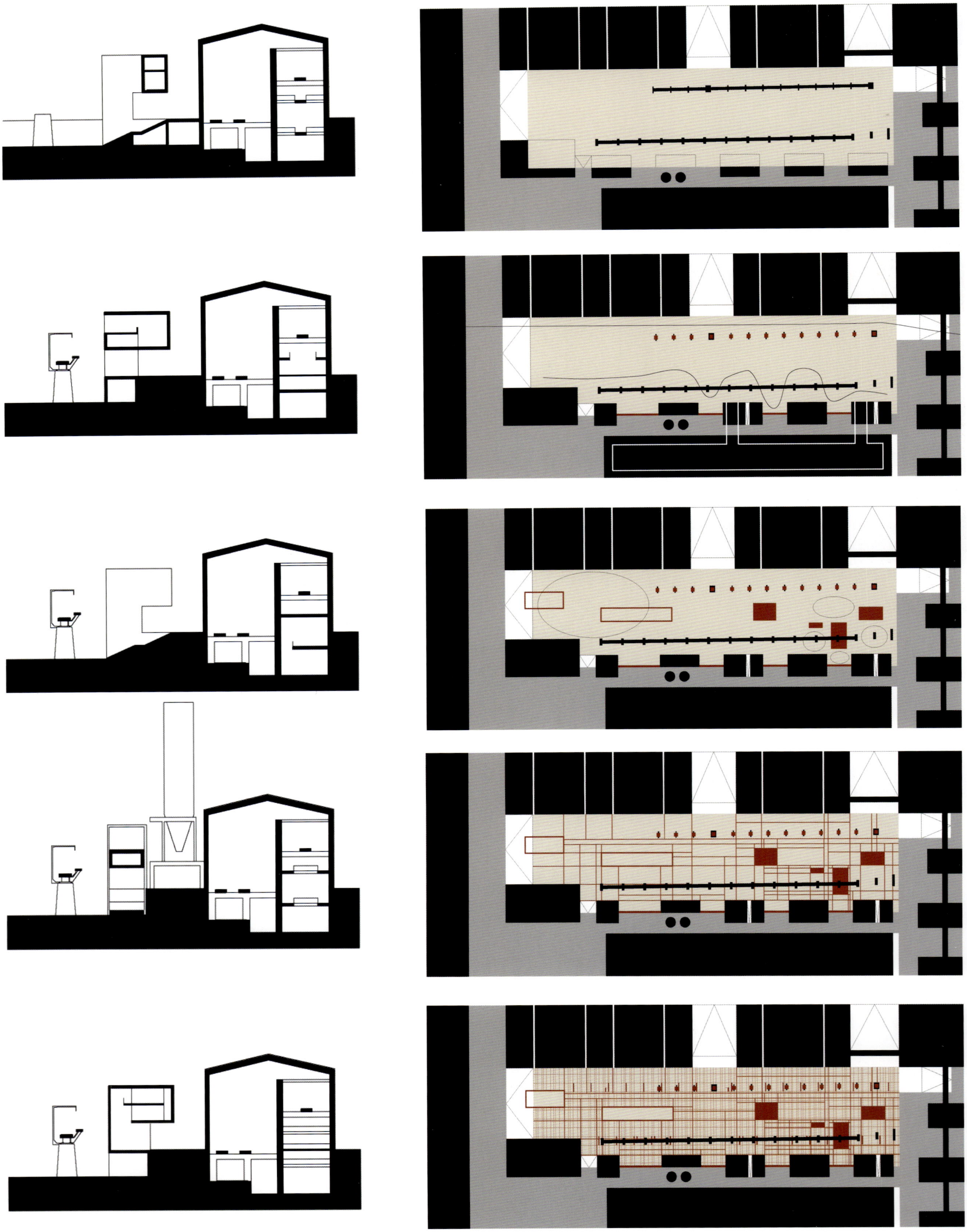

2012年6月26日，建于Esch-sur-Alzette的一个钢厂旧址上的卢克斯广场项目在极富盛名的卢森堡公国建筑设计比赛中获得了空间与景观类大奖，阿格拉和奥尔威古特事务所在卢森堡公国举办的节日庆典中接受了这一奖项。

改造之前，废弃钢厂的旧址占地广阔、环境荒凉，地面上遍布着一些苔藓类植物和零星桦树。奥尔威古特事务所并没有完全摈弃原有的一切，而是以一种较为温和的方式，将原有地块中某些隐藏起来的特点保留下来。设计师们将那些历经风雨的建筑材料，如混凝土、实木以及未经加工的钢材等完全融入到这个类似工厂厂房的建筑设计之中。空空的座椅、台阶和绿化景观相互交错，形成一种特别的空间视觉效果。这些古旧材料上面锈迹斑斑，刻满了时光留下的痕迹，展现了一种时光流逝，历史重现的感觉。

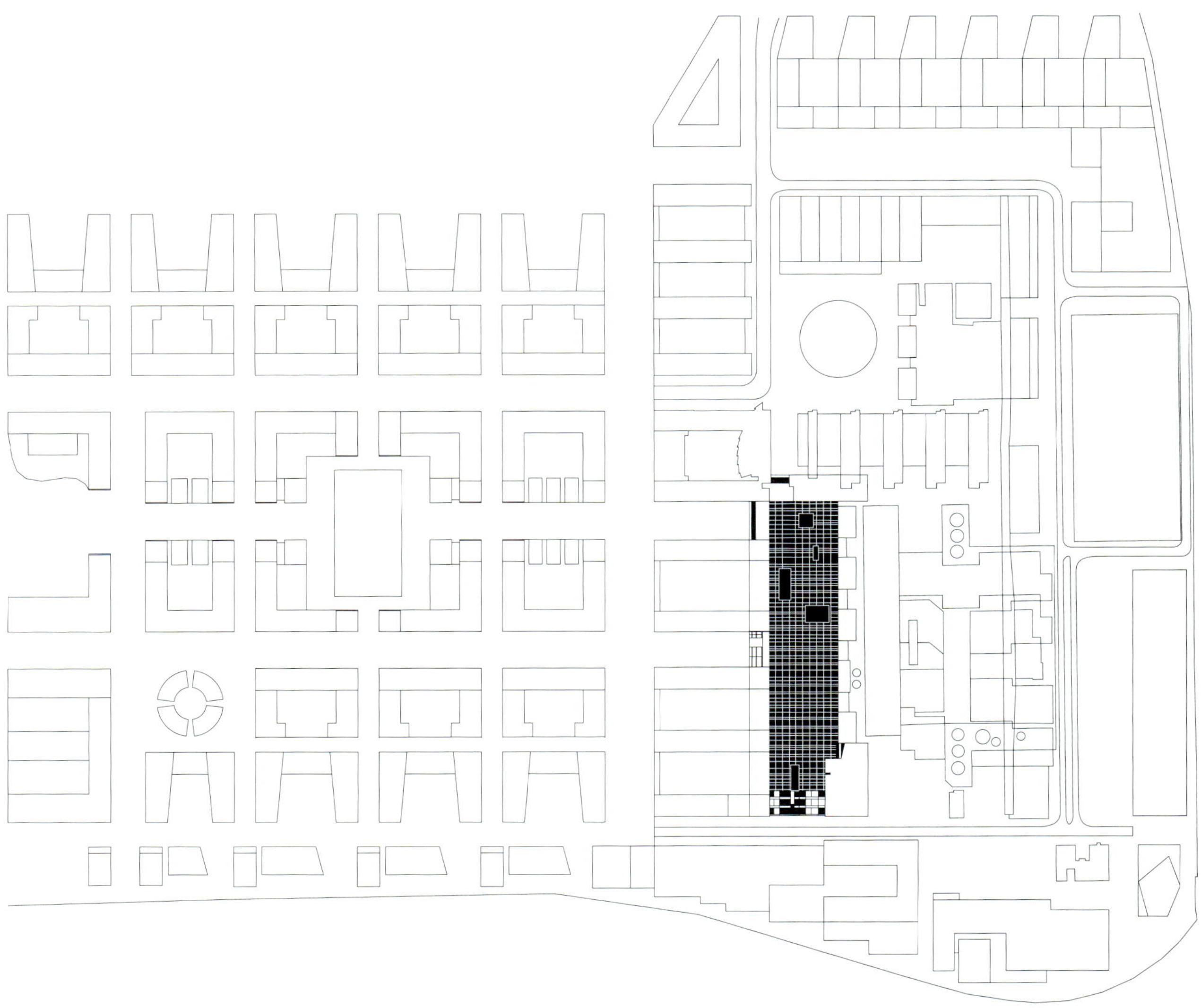

LUX

ANNETTE THOMMESSENS SQUARE

安妮特托马森广场

Oslo, Norway

Annette Thommessens Square is named after the first head of the Norwegian Organization for Asylum Seekers (NOAS).

The pedestrian bridge crossing the railway tracks of Oslo Central Station ends at Annette Thommessens Square.

The bridge is the main pedestrian link between the inner city east and the new urban renewal waterfront of Bjørvika.

The shape of the square has been designed to accommodate its function as a hub with a heavy flow of pedestrian traffic, as well to provide an attractive meeting venue.

The entire square catches the sun all day, making it attractive as a social meeting place. At night the square is lit by four floodlights mounted on the angled light mast and is supplemented by up-lights embedded in the granite tiles.

Columnar English oak trees have been planted in a row beside the outdoor café in front of the Tax Building.

The square has universal accessibility. All slopes are gentle (less that 1:20) and the pedestrian bridge can be reached via stairs, lift or the ramp. There are plentiful accesses to three bike stands distributed over around the square.

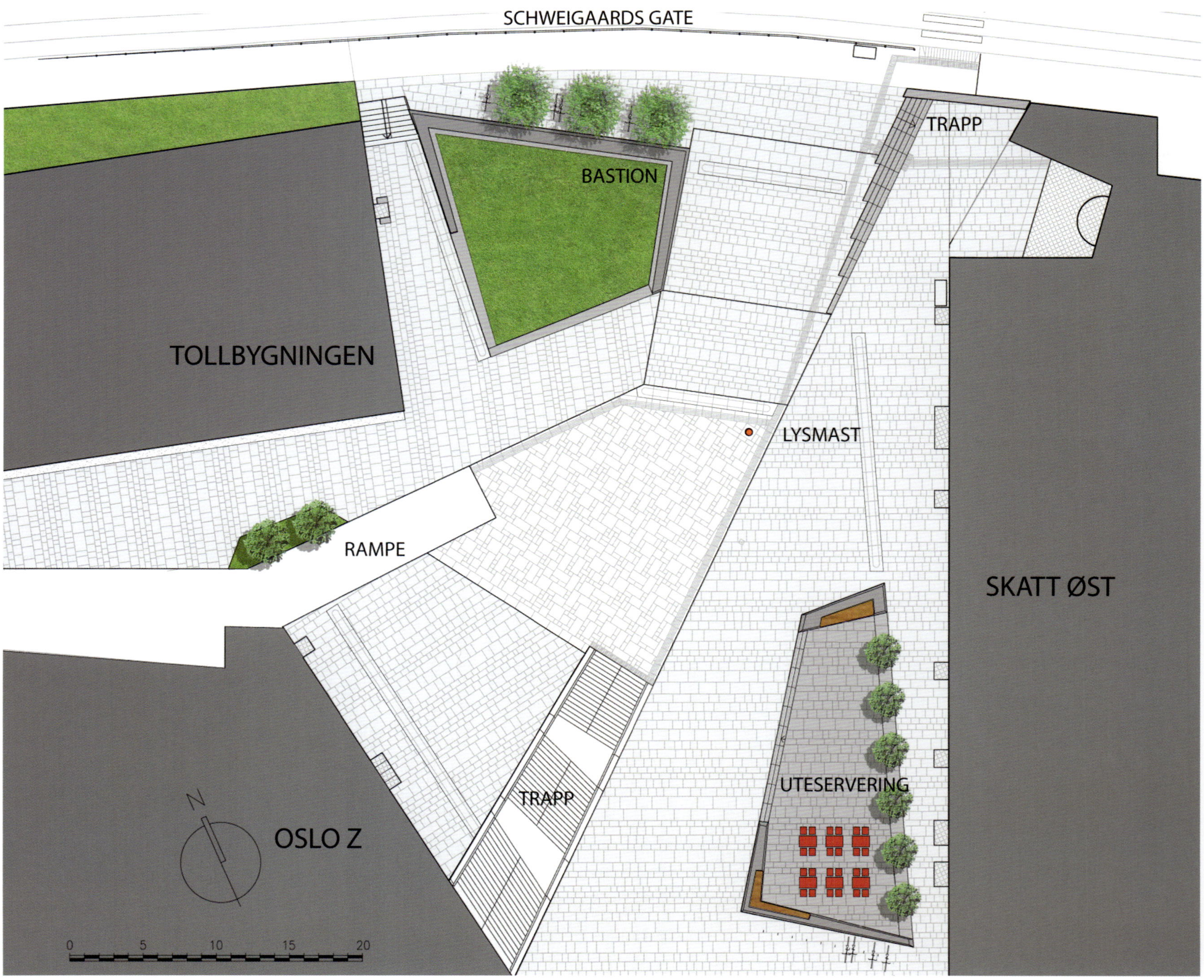

ØST
Schweigaards gate 15
POLITI

annette thommessens plass

安妮特托马森广场是以挪威寻求庇护者组织（NOAS）第一任领导人的名字命名的。

人行天桥横跨奥斯陆火车站的铁路线，直通安妮特托马森广场。

人行天桥是连接城市东部和重建的新都市区——比约溪滨水地区之间主要的人行通道。

广场的形状是根据其功能设计完成的。这里将成为一个行人流量极大的交通枢纽，也可以根据需要改造成一个引人注目的会议场地。

整个广场白天阳光充足，是一个极具吸引力的公众社交场合。晚上则灯火通明，安装在广场四周倾斜灯杆上的四盏泛光灯是主要光源，嵌在花岗岩砖里的灯光射向上方作为补充。

税务大楼前方种植了一排柱状英国橡树，再往前是一个户外咖啡馆。

整个广场交通便利，周边所有的坡道坡度很小（斜率小于1:20），公众可以通过电梯、楼梯或者坡道直接通往人行天桥。另外，还设有众多的通道通往分布在广场附近的三个自行车停放站。

• **Landscape Architect** - György Szloszjár, Borbála Salgó, Eszter Dandé • **Firm** - Garten Studio Ltd.

ÚJBUDA (NEW BUDA) CITY CENTER

布达佩斯新城市中心

Budapest, Hungary

According to the designers' motto, the image of a public space – apart from the indispensable functionality – is determined by original solutions of forms and by the spaces of interesting personal experience and action. This is why they put special emphasis on the uniqueness of the environment. Their aspiration in this regard may be recognized in the architecture of claddings, in the illumination of open spaces or the appearance of the water feature, the unique design of the playground fence and the street furniture.

The main concept of the whole new block was a fluent inside with a structured outside facade, in various textures. The aim was to adapt the buildings in the former city structure and scale. So in outside next to the new promenades there are smaller "micro-spaces" with their own formal language too. After a year of active usage they experienced that the functional units are working well. The office blocks and flats are oriented to the narrow part of the street with less public spaces, while the multifunctional shopping center opens up towards the surroundings, even the outside pavement pattern continues in the inside. In the evening the landscape structures are underlined with lighting elements too. Material usage of the streetscapes is a bit different from each other but there are always common materials like limestone and common street furniture. In big units they used basalt, dark brick, and an exclusive granite top concrete pavement, while the stripes are usually made from limestone. There is unique street furniture on each promenade, like the hardwood covered concrete benches on air inlets, or the curved steel benches in pairs forming tree guards.

• **Area** - 20,000 m^2 • **photographer** - Attila Polgár, Garten Studio Ltd.

vízmedence
vízmedence
vízmedence

VÁSÁRCSARNOK
CINEMA CITY
ELECTRO WORLD
INTERSPAR
BUSZ

ÚJ SKÁLA ÉPÜLETE
PV ±0,00=105,50 mBf

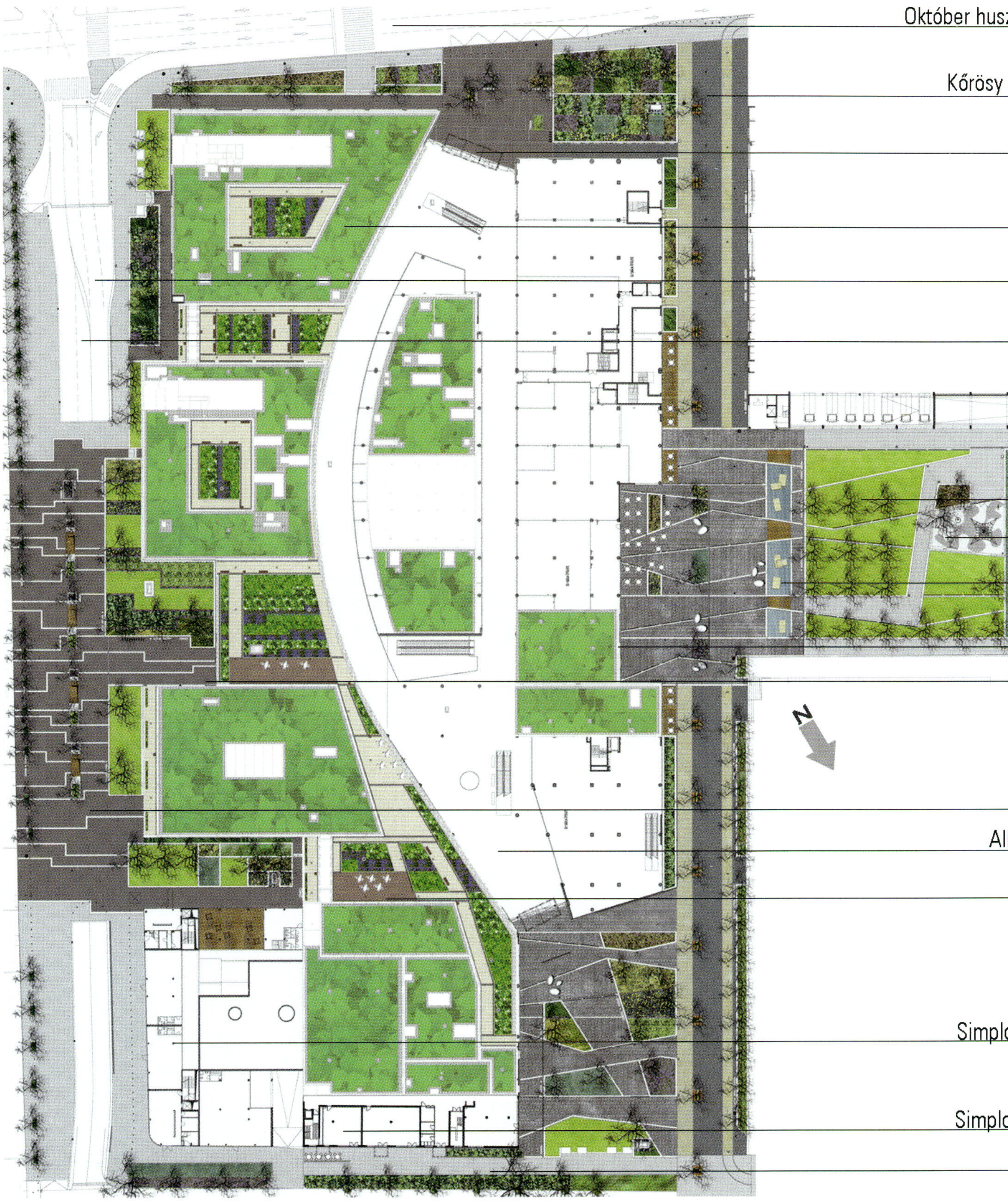

布达佩斯新城市中心的设计师们遵循这样一个理念：对于公共空间来说，除了必不可少的功能空间和设施之外，其他部分应该优先考虑最基本的需求和民众的意愿和感受。因此，设计师们将设计规划的焦点集中在如何突出和强调广场项目独一无二的环境特点。这种环境特点最终体现在重叠式的建筑风格、广场空间的照明体系、精心设计的水系景观、匠心独具的广场围墙以及遍布街头的不同景物上。

广场项目规划的主要目的是利用不同的结构，结合内部流畅的线条和外部不同的建筑风格，打造一个崭新的城市中心，同时，新建的建筑自然地融入原有的城市结构和规模建筑体系当中。有鉴于此，在广场外围紧邻新建的人行步道旁，那些更小的“微空间”都保持了原有的形态。在广场项目建成并对外开放一年后，这些“微空间”都很好地起到了功能空间单位的作用，运行良好。广场项目管理办公室朝向一条狭窄的街道，占地面积不大，多功能的商业中心直接通往周边地区，甚至设有专门的人行步道从商业中心一直延伸到广场外侧。入夜，广场景观结构在灯光体系的映衬下分外耀眼。制作每一幅街景画的材料都不尽相同，但大多都采用类似石灰石等普通街景材料。在一些较大规模的空间里使用了玄武岩、暗色砖块和等级颇高的花岗岩铺设混凝土路面，甚至条纹都是利用石灰石材料构筑的。每条步行街区都具有各自独特的街景，如在通风口上方利用硬木材料包裹的混凝土长椅，又如利用成对的曲线形钢制长椅作为树木的护栏等。

HUMANIC
VAN GRAAF

ZARA
CINEMA CITY
ELECTRO WORLD
INTERSPAR
allee

FITNESS
HIPER MARKET
1200 Parkolóhely
Beragyogja minden napját.
Allee.
allee.hu
Coca-Cola

150
ÜZLET
FITNESS
HIPER
MARKET
1200
Beragyogja
minden napját.
Allee.

5+
6+
10+

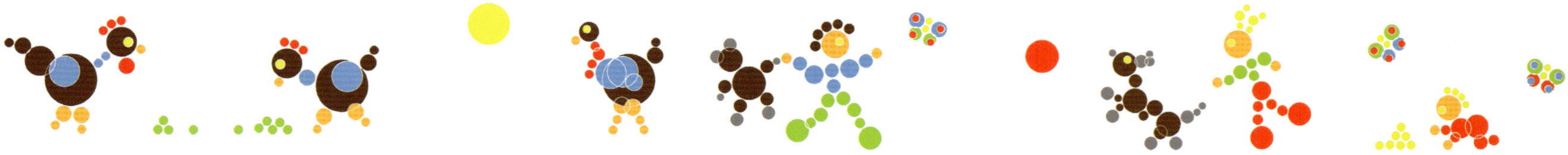

promod

ZARA

MILLENNIUM CITY CENTER

Budapest, Hungary

布达佩斯千年城市中心

The main concept of the whole new block was forming a fluent ground floor with public functions, involving the surrounding open space into it. Another intention on higher levels was to divide the longish shape, making structural and visual segmentation and let through the pedestrians cross traffic towards the river and the promenade.

The designers mixed a contemporary design of triangle shape surface forming, with a little origami, which both gave them enough freedom to make the functional spaces and get the unusual, constructed look from above.

The complete project is expected to earn LEED Gold rating, which had effects on design process too. Mostly the adopted technology of drainage, irrigation and lighting and material usage of paving materials and plantation had to be considered by LEED requirements.

Every member of the design team has his own personal favorite detail. They abounded in designing water features – there are four on the site. Two of them are next to the building's coffee terraces, and the other two are parts of the folded terrain. Each of them has its own specialty: M1 – its water basin was made from huge carved white granite elements; M2 has a pixel effect of "designer's" handprint on the spot, and was made from split granite stripes, combined with flowing water and steam jets; M3 has a sloped, plus to minus basin surface with flowing and stagnant water; M4 was formed with in level water surface and its vivid lights were combined with water jets. There are lots of invisible technologies underground connected to water features and drainage system. The lightness of the only, 8 cm thick retaining walls next to the folded grass slopes could be achieved with a mounted top stone. The wavy walkway is covered with the same sized trapezoid wood boards, and their directions rotate depending on the size of the loop.

• **Area** - 7,000 m^2 • **photographer** - Garten Studio Ltd., István Steffler

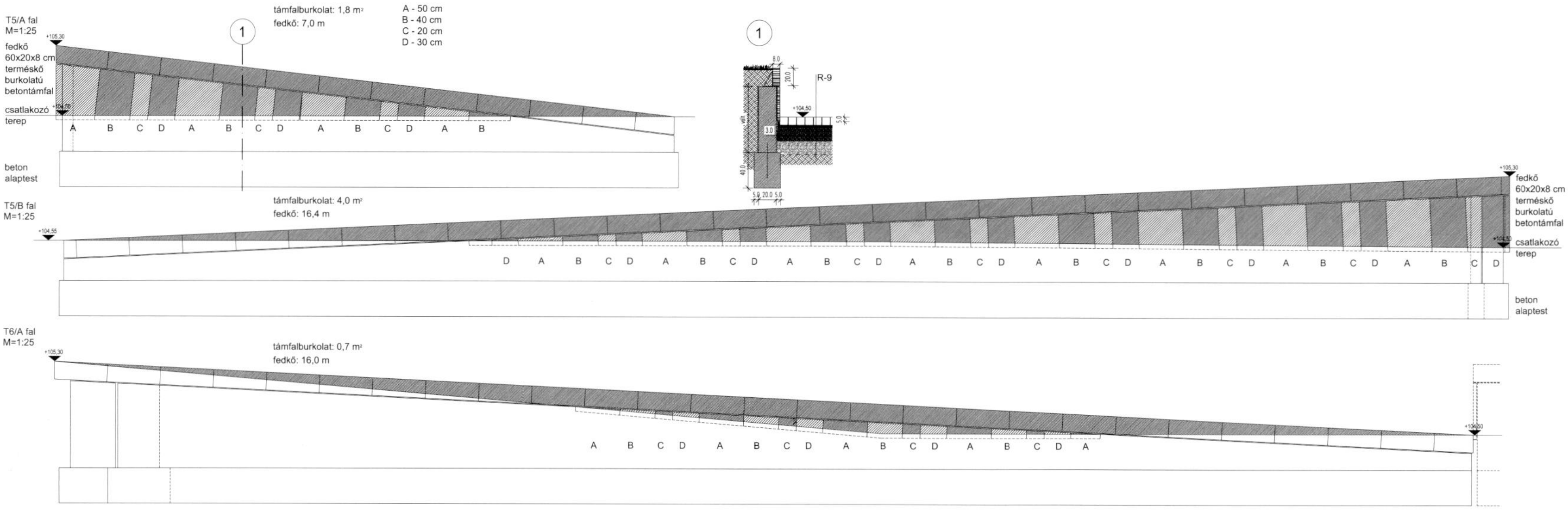

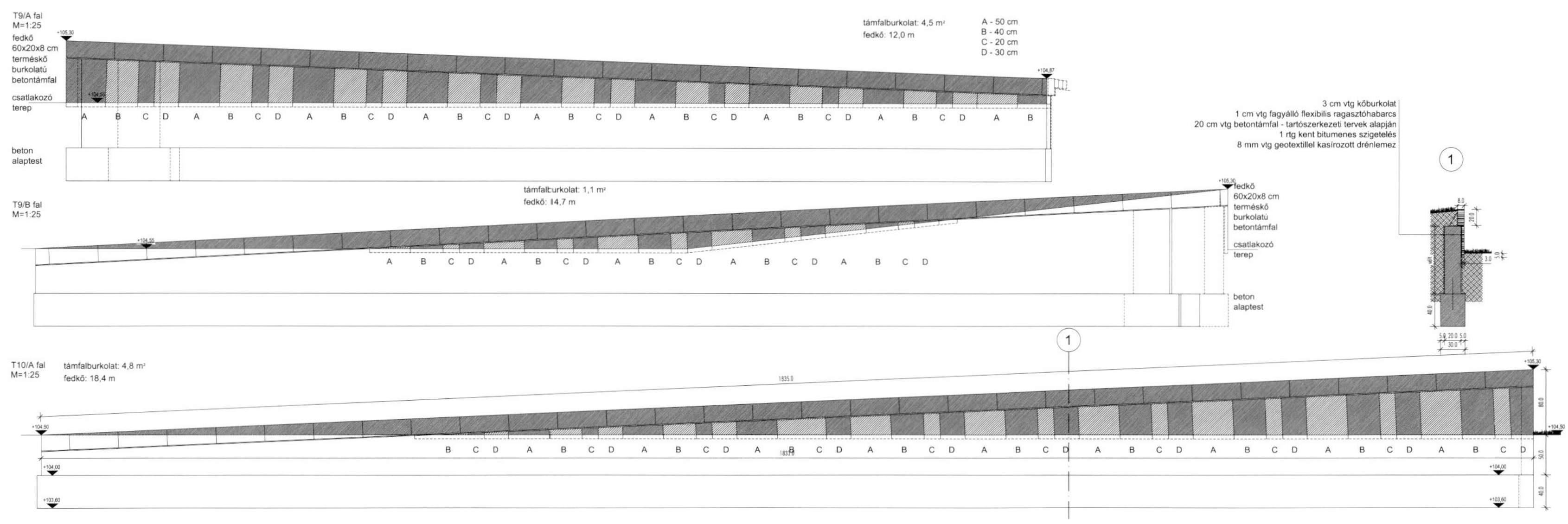

M2 MEDENCEÉHEZ CSATLAKOZÓ VILÁGOSSZÜRKE TERMÉSZETES KŐ GYALOGOSBURKOLAT KIOSZTÁSA
M=1:50

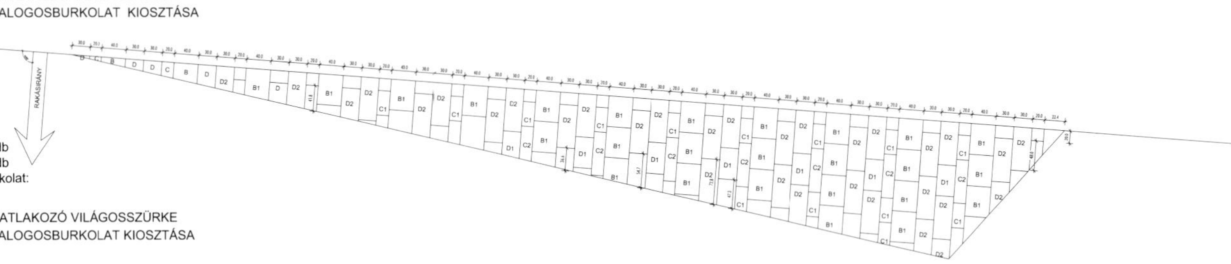

B1 - 40x50 cm - 31 db
C1 - 20x40 cm - 18 db
C2 - 20x70 cm - 10 db
D1 - 30x50 cm - 15 db
D2 - 30x70 cm - 31 db
20x20 cm kezdőkő - 12 db
20x30 cm kezdőkő - 11 db
M2 medence melletti burkolat:
18,19 m²

M3 MEDENCEÉHEZ CSATLAKOZÓ VILÁGOSSZÜRKE TERMÉSZETES KŐ GYALOGOSBURKOLAT KIOSZTÁSA
M=1:50

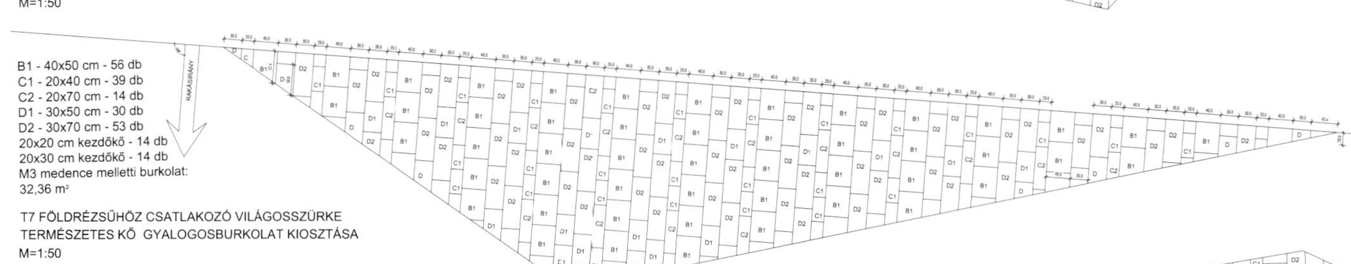

B1 - 40x50 cm - 56 db
C1 - 20x40 cm - 39 db
C2 - 20x70 cm - 14 db
D1 - 30x50 cm - 30 db
D2 - 30x70 cm - 53 db
20x20 cm kezdőkő - 14 db
20x30 cm kezdőkő - 14 db
M3 medence melletti burkolat:
32,36 m²

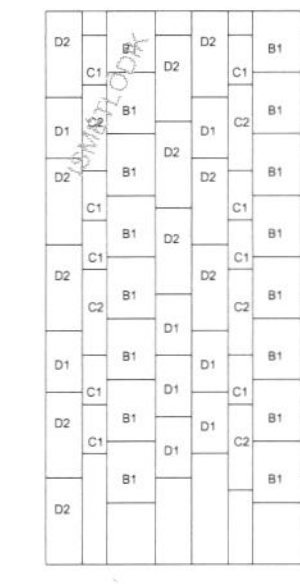

T7 FÖLDRÉZSŰHÖZ CSATLAKOZÓ VILÁGOSSZÜRKE TERMÉSZETES KŐ GYALOGOSBURKOLAT KIOSZTÁSA
M=1:50

B1 - 40x50 cm - 60 db
C1 - 20x40 cm - 45 db
C2 - 20x70 cm - 22 db
D1 - 30x50 cm - 74 db
D2 - 30x70 cm - 39 db
30x40 cm kezdőkő - 20 db
20x30 cm kezdőkő - 21 db
10x20 cm kezdőkő - 20 db
T7 földrézsü melletti burkolat:
41,47 m²

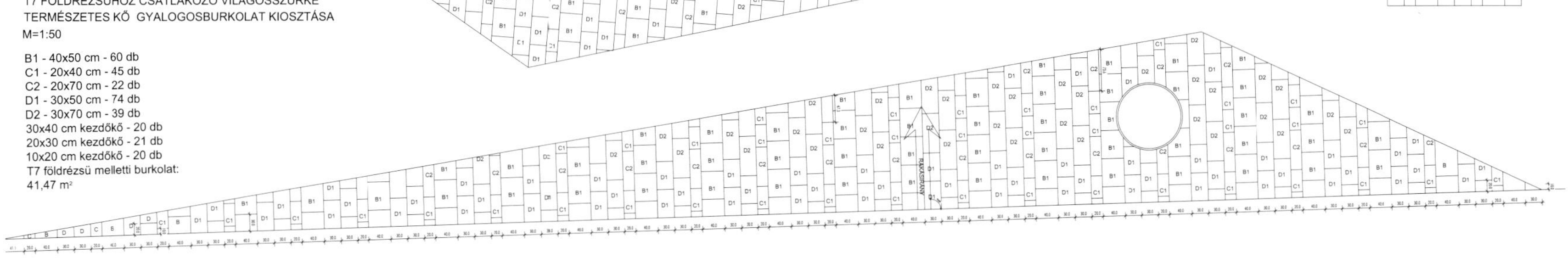

M1

千年城市广场项目的主要理念是建成一个平滑的广场地面，为公众打造一个具有公共功能的空间，与周边的环境融为一体，并利用步行通道将原有的长条形空间进行分割，形成更舒适的结构和视觉效果，这样，游客们可以沿着步行通道直达河边。

设计师们在项目的设计规划中混合了现代设计风格中的三角构形概念，加入少许折纸艺术风格，给广场留下了足够的功能空间，并获得了独特的设计效果。

建成后的广场项目有望获得 LEED 黄金等级认证，这一认证影响着设计的整个过程。在排水、灌溉和灯光体系中运用的科学技术以及在地面铺设和景观绿化方面的材料选择，都达到了 LEED 所规定的标准。

广场项目设计组的每名成员都力争在设计细节中体现各自的特点，他们在项目设计规划中加入了四个特点鲜明的水系景观（1-4 号），其中两个位于建筑平台上的咖啡屋旁边，另外两个位于具有折纸造型的区域内。每个水景设计都特色鲜明：1 号景观是一个池塘，周边饰有巨大的花岗岩石雕；2 号利用了设计师的手印造型，表现出一种抽象的视觉效果；3 号略带坡度，流水潺潺；4 号利用灯光和水平落差，形成了一个五光十色的喷泉。当然，在水系景观和地下排水系统的工程中还运用了大量不为人知的技术手段。折纸造型的玻璃斜坡旁边是轻型挡土墙，利用顶级石材砌成，仅有 8 厘米厚。起伏的步行路面由大小相同的不规则四边形木板铺设而成，根据坡度的大小来改变铺设方向。

M1

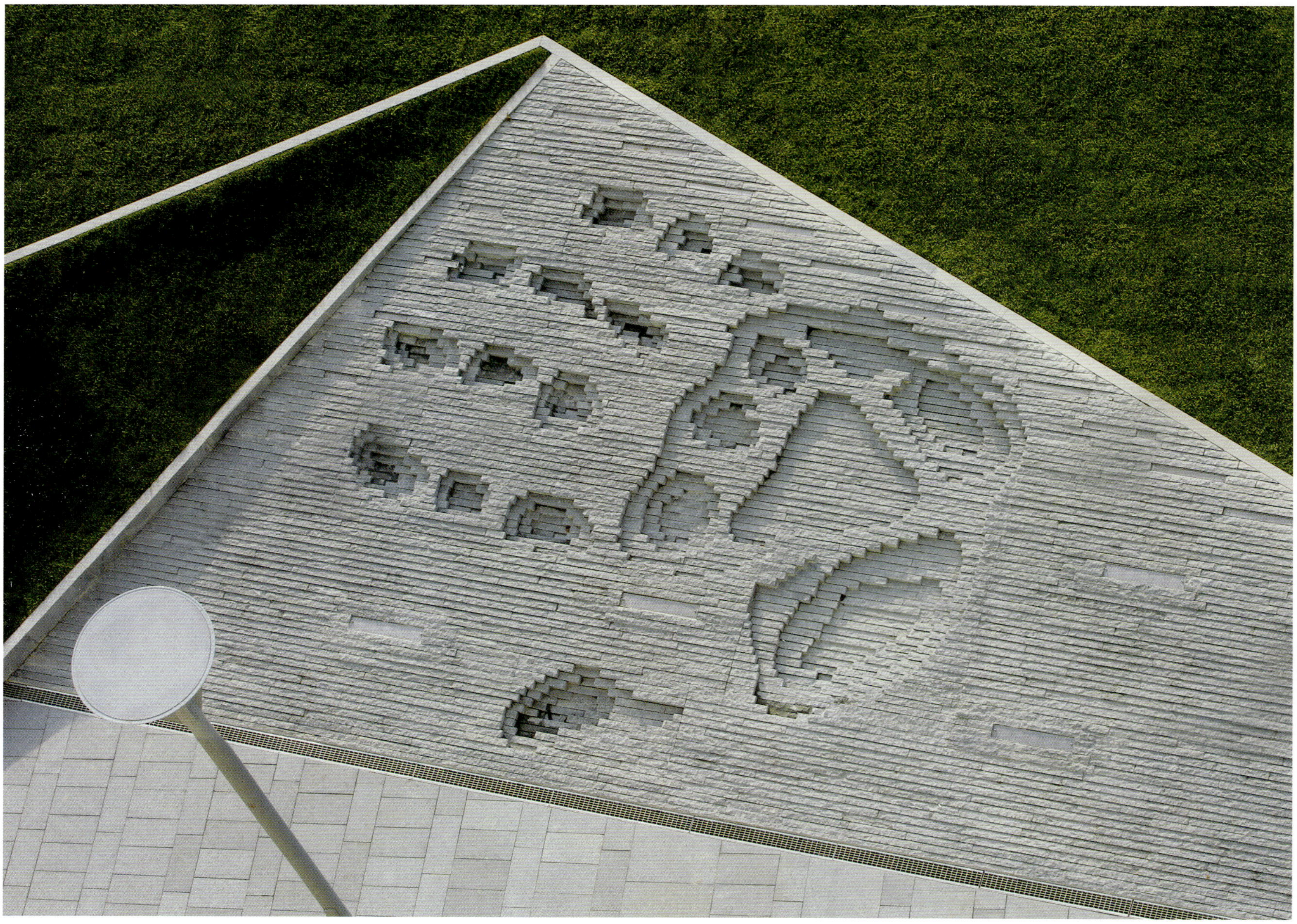

M2

M2

M2-ES MEDENCE
M=1:25

MEDENCE ÉLÉRE ÁLLÍTOTT KŐBURKOLATÁNAK MENNYISÉGE:

ÉLÉRE ÁLLÍTOTT VILÁGOSSZÜRKE KŐBURKOLAT
FELSŐ ÉLE MEGMUNKÁLT (hasított):

5x5 cm keresztmetszet: 27,9 m² - 558 fm ebből levonódik a rozsdamentes fúvókák hossza 15,1 fm összesen 558-15,1=542,9 fm

ÉLÉRE ÁLLÍTOTT VILÁGOSSZÜRKE KŐBURKOLAT,
VÁLTOZÓ VASTAGSÁGBAN PLASZTIKA TERÜLETÉN
FELSŐ ÉLE MEGMUNKÁLT (hasított):

TEREPPLASZTIKA:
- 18x5 cm keresztmetszet: 4,6 m² - 92 fm
- 13x5 cm keresztmetszet: 3,4 m² - 68 fm
- 8x5 cm keresztmetszet: 2,9 m² - 58 fm
- 5x5 cm keresztmetszet: 1,0 m² - 20 fm

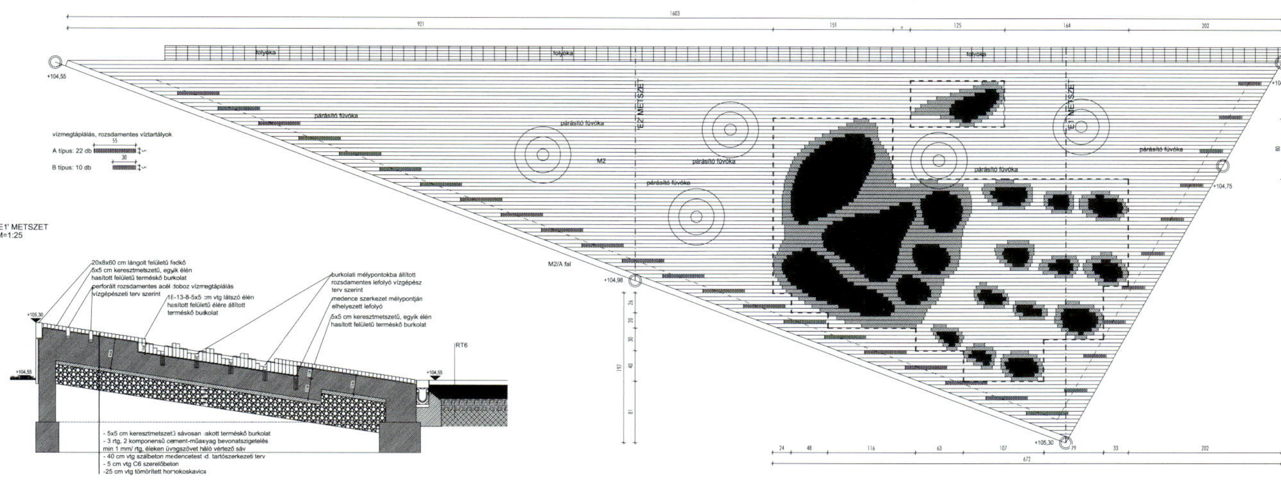

'E2' METSZET
M=1:25

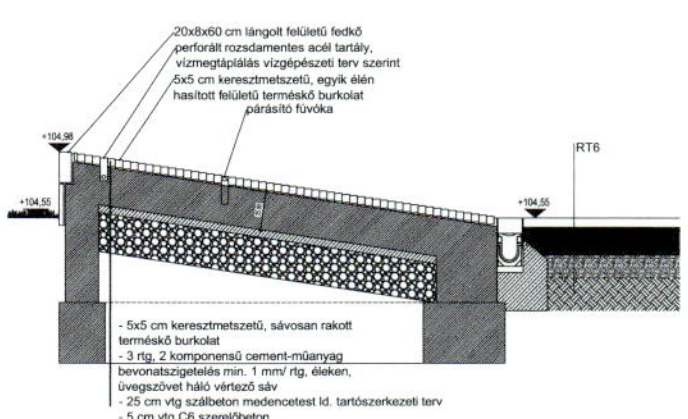

'E1' METSZET
M=1:25

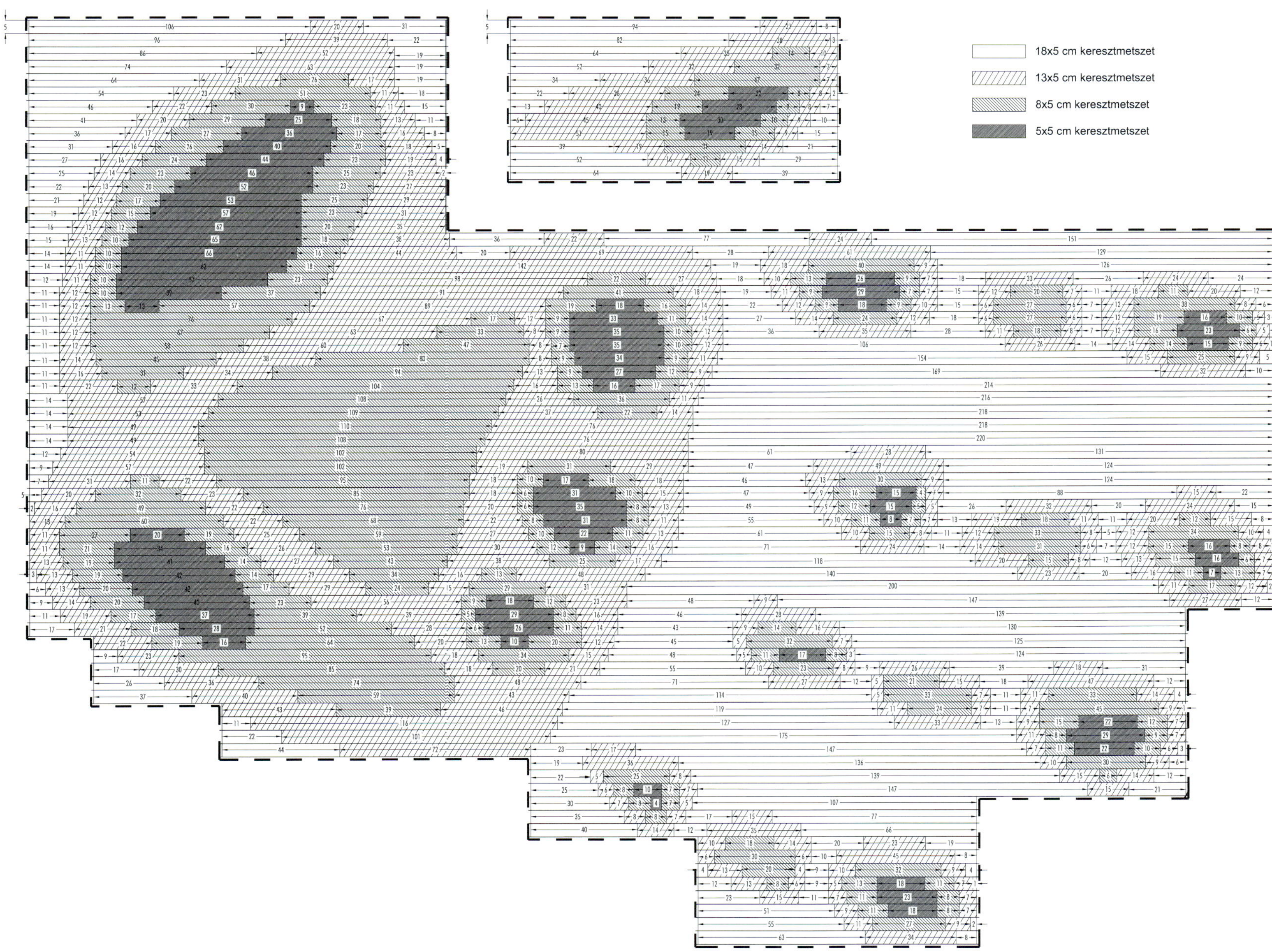

- 18x5 cm keresztmetszet
- 13x5 cm keresztmetszet
- 8x5 cm keresztmetszet
- 5x5 cm keresztmetszet

M3

M4

- **Architect** - Malaka Ackaoui
- **Firm** - Williams, Asselin, Ackaoui & Associates Inc.

JAMES SQUARE

詹姆斯广场

Canada

James Square is a quintessential 21st century design that reinvents campus space to fit contemporary life and extends educational activities into the landscape in a sustainable manner. It highlights McGill University's commitment to improve its environment and act as a responsible institution. The project was lead by WAA in collaboration with Genivar, EGP, BPR & CS Design.

The design reflects current educational and social needs while respecting the historic heritage of McGill University's buildings in the 19th century. Clearly, contemporary gestures complement and set the standard for sustainable practices in campus design.

James Square's redesign resulted in a shift from a vehicle-dominated space to a pedestrian and bicycle focused zone. Many seating and socializing opportunities were created by treating the topography as an asset rather than an obstacle and five sculptures add to the square's cultural landscape. Increased soft-surfaces and the use of indigenous planting absorb surface run-off water and offer a refuge to wildlife.

James Square is a cultural site where landscaping, nature, art and engineering work together to provide McGill campus users a chance to fully profit from their on-campus experience. Place-making transformed the square from a circulation interchange to a space where social activities and academic functions mingle in the core of downtown Montreal.

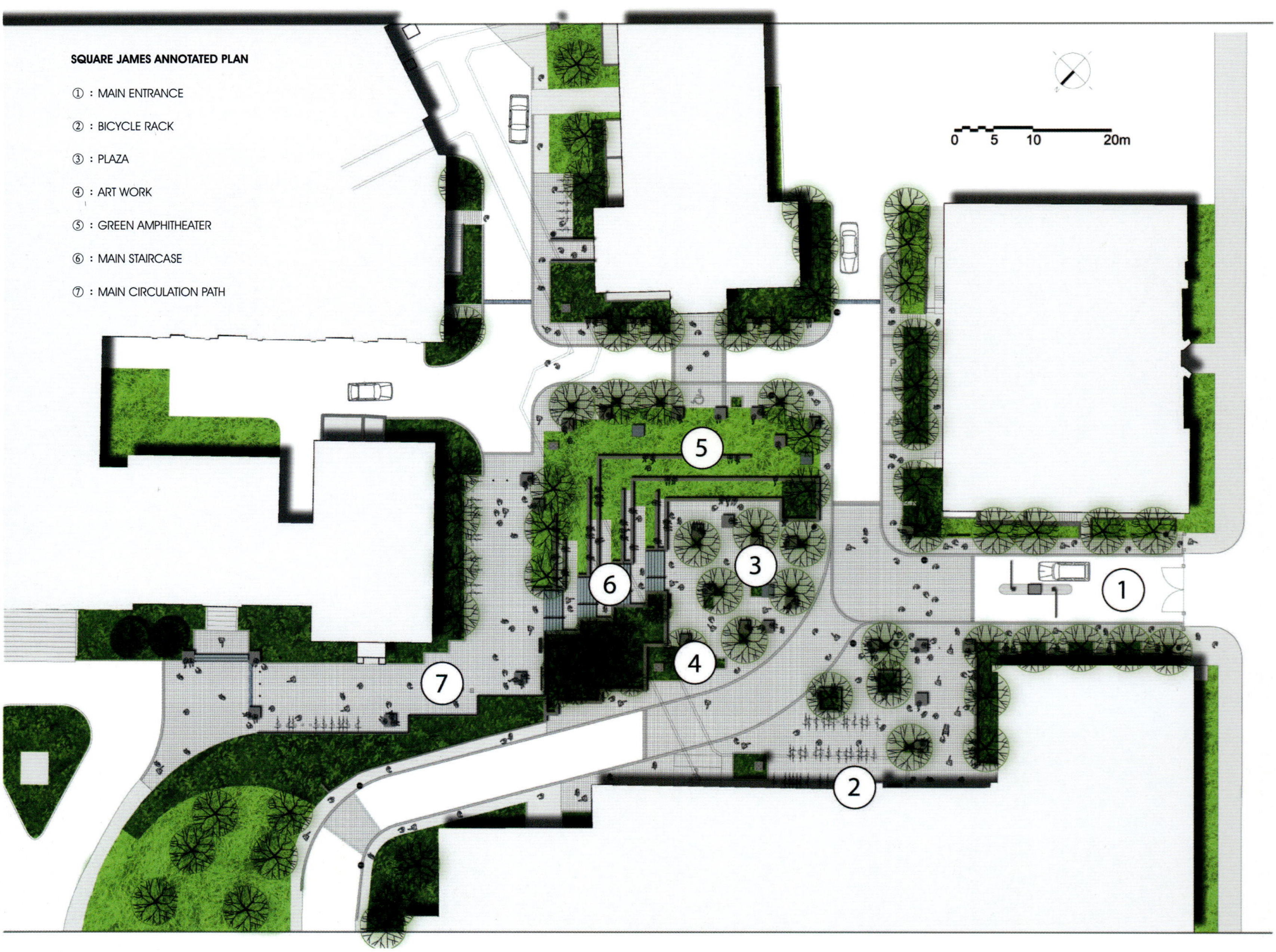

- **Area** - 6,000 m²
- **Photographer** - Vincent Asselin and Groupe TREMCA Inc.

James

詹姆斯广场是一个典型的21世纪的设计，重塑了校园空间以适应当代生活，并以可持续发展的方式将教育活动延伸到景观之中。它突出了麦吉尔大学的承诺：改善自身环境，成为一所富有责任感的院校。该项目由WAA组织领导，并与Genivar、EGP、BPR & CS Design合作完成。

设计反映了当前的教育和社会需求，同时也尊重了麦吉尔大学19世纪建筑的历史遗产。很显然，本项目在校园设计方面使用了当代建筑手法进行补充，并采用了可持续发展的标准。

詹姆斯广场的重新设计，将原来机动车占主导地位的空间转变成注重行人和自行车活动的区域。地势的问题没有被视为障碍，而是作为优势为人们创造了许多休息和社交的机会。同时，五处雕塑还增添了广场的文化景观。软地面的增设和本土植物的使用吸收了地表的径流水，也为野生动物提供了栖息的场所。

詹姆斯广场是一个文化遗址，集景观、自然、艺术和工程设计于一体，为到访麦吉尔大学校园的人们提供了充分享受校园体验的机会。经过改造，该广场从一个仅仅具有交通功能的地方，变成了一个位于蒙特利尔市中心的社会活动和学术活动交融的核心区域。

AMSTELVEEN ZONNEHUIS CARE HOME

Amstelveen, The Netherlands

阿姆斯特尔芬综合关怀中心

Town dwellers need high-quality green space to enhance their living experience. HOSPER has produced a proposal for development of the grounds of the Zonnehuis care home in Amstelveen into a lively green pedestrianised area. The various buildings on the site are linked by a square paved in a distinctive design housing a number of richly planted green oases and play area and other facilities. Four residential units on a multifunctional plinth, designed by Rijnboutt, will be erected on the west side. A water-rich patio garden has been designed as part of this complex.

The new Zonnehuis care complex, designed by Architectuurcentrale Thijs Asselbergs, has already been completed on the east side. The outside space surrounding the care complex consists of gardens with perennial plants and a greenhouse where residents can garden. The square includes a terrace, a play area, a garden area with perennial plants and a sloping path and steps for walking exercises. Two gardens have been specially designed to be dementia-friendly, with the aid of advice from Anke Wijnja of Bureau Fonkel and Annie Pollock of the Dementia Services Development Center at the University of Stirling in Scotland. The two halves of the care complex are connected by a covered pergola (which can still be used in case of rain) designed jointly by Octatube, Architectuurcentrale Thijs Asselbergs and HOSPER.

One special feature is that the entire grounds of the care home (with the exception of the dementia-friendly gardens) are freely accessible to residents of the surrounding neighbourhood.

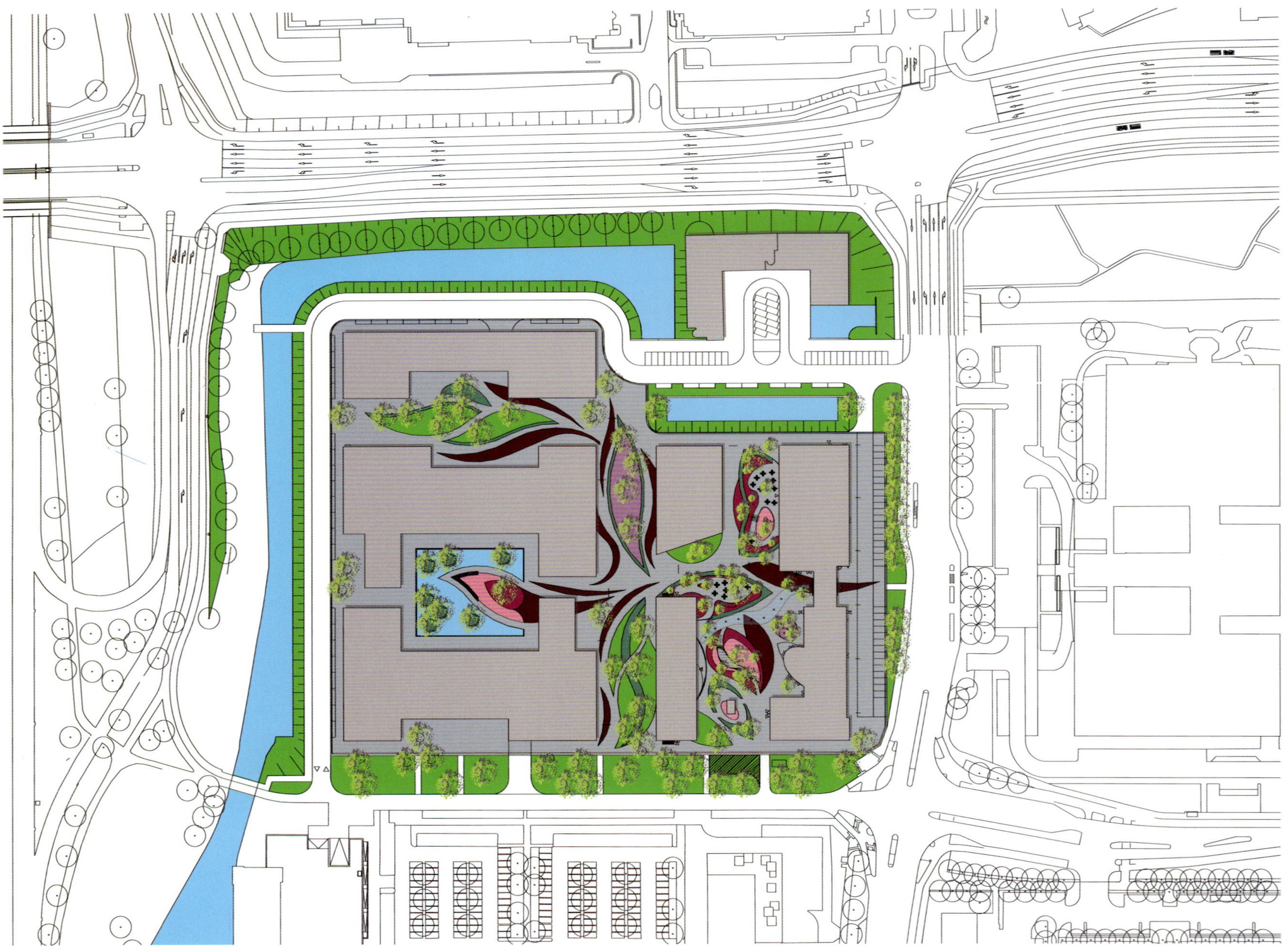

• **Partner** - DG groep,Bureau Fonkel, Dementia Services Development Centre, Rijnboutt, Architectuurcentrale Thijs Asselbergs, Octatube

• **Client** - Zonnehuisgroep Amstelland Foundation, M.J. de Nijs projectdevelopment

• **Area** - 30,000 m^2

• **Photographer** - Pieter Kers

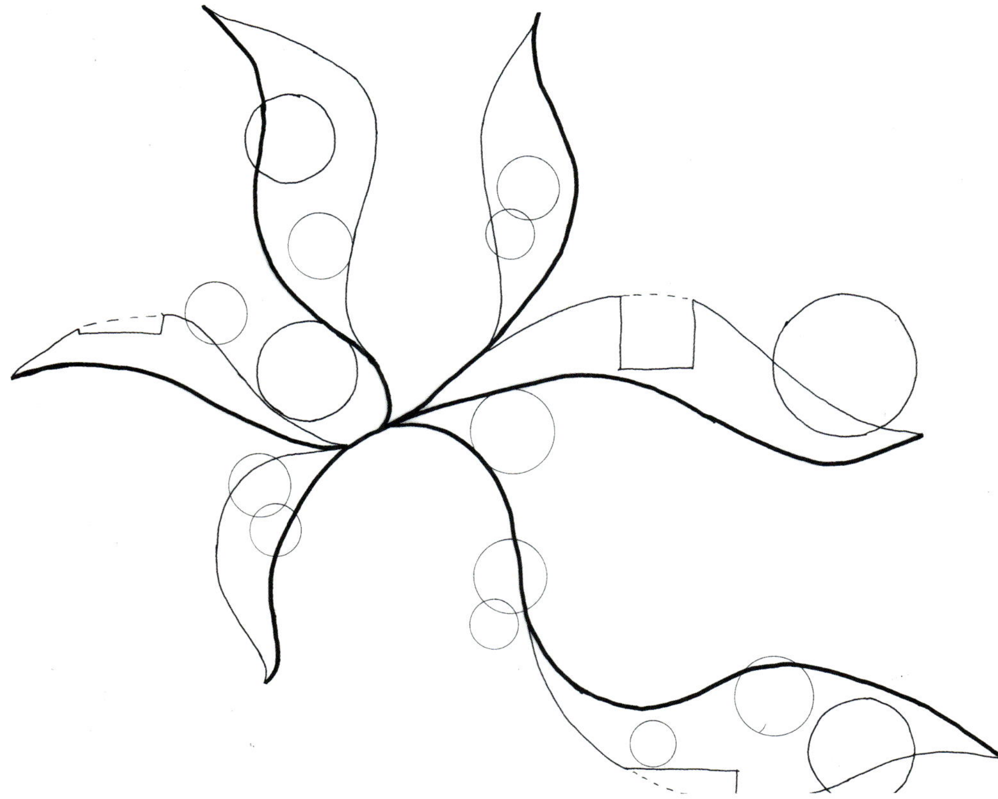

阿姆斯特尔芬综合关怀中心项目的实施旨在为城市居民提供高质量的绿色区域，提高居民的生活质量。HOSPER 提议将这一区域建设成一个生机勃勃的绿色步行区域。区域内不同的建筑通过一个广场相互连通。广场的设计独具特色，包括多个绿色区域和公共娱乐设施。位于西侧的多功能区域建有四个居住单元，由设计师莱茵布特负责设计，旁边还设有一个带有水系景观的花园。

新的关怀中心建筑位于东侧，由西斯·阿瑟伯格设计中心负责设计规划，现在已经完工。关怀中心周边是比较开阔的户外空间，花园里种植着多年生植物，还建有一个温室，居民可以在里面种植想要的植物。建筑周边的广场里设有一个露台、一个娱乐场、一个多年生植物园和供行人行走的斜坡和台阶。其中两个花园由 Fonkel 机构的安可·维吉尼亚和苏格兰斯特林大学智障服务发展中心的安妮·波洛克提议，专门为智障患者设计的活动场所。关怀中心的两部分建筑将由 Octatube、西斯·阿瑟伯格设计中心和 HOSPER 共同设计完成的绿廊连接（雨天可兼作遮雨棚）。

在该项目中，除了为智障患者专门设置的花园区域外，整个区域都免费向周边居民开放。

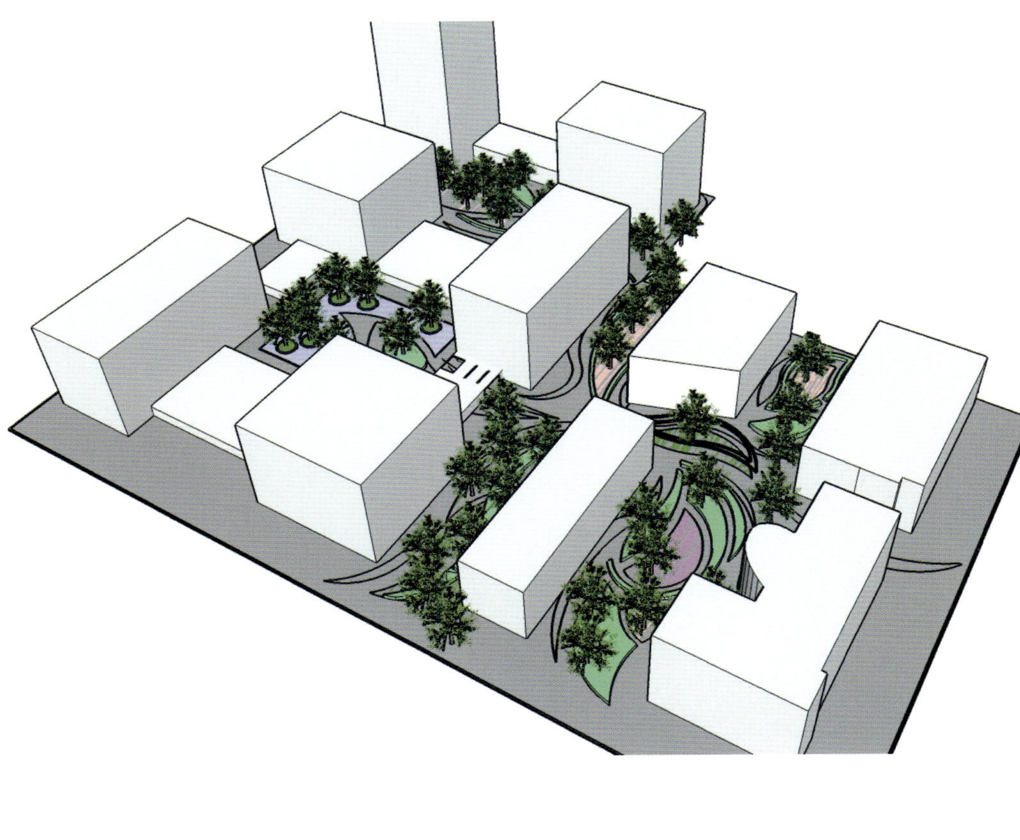

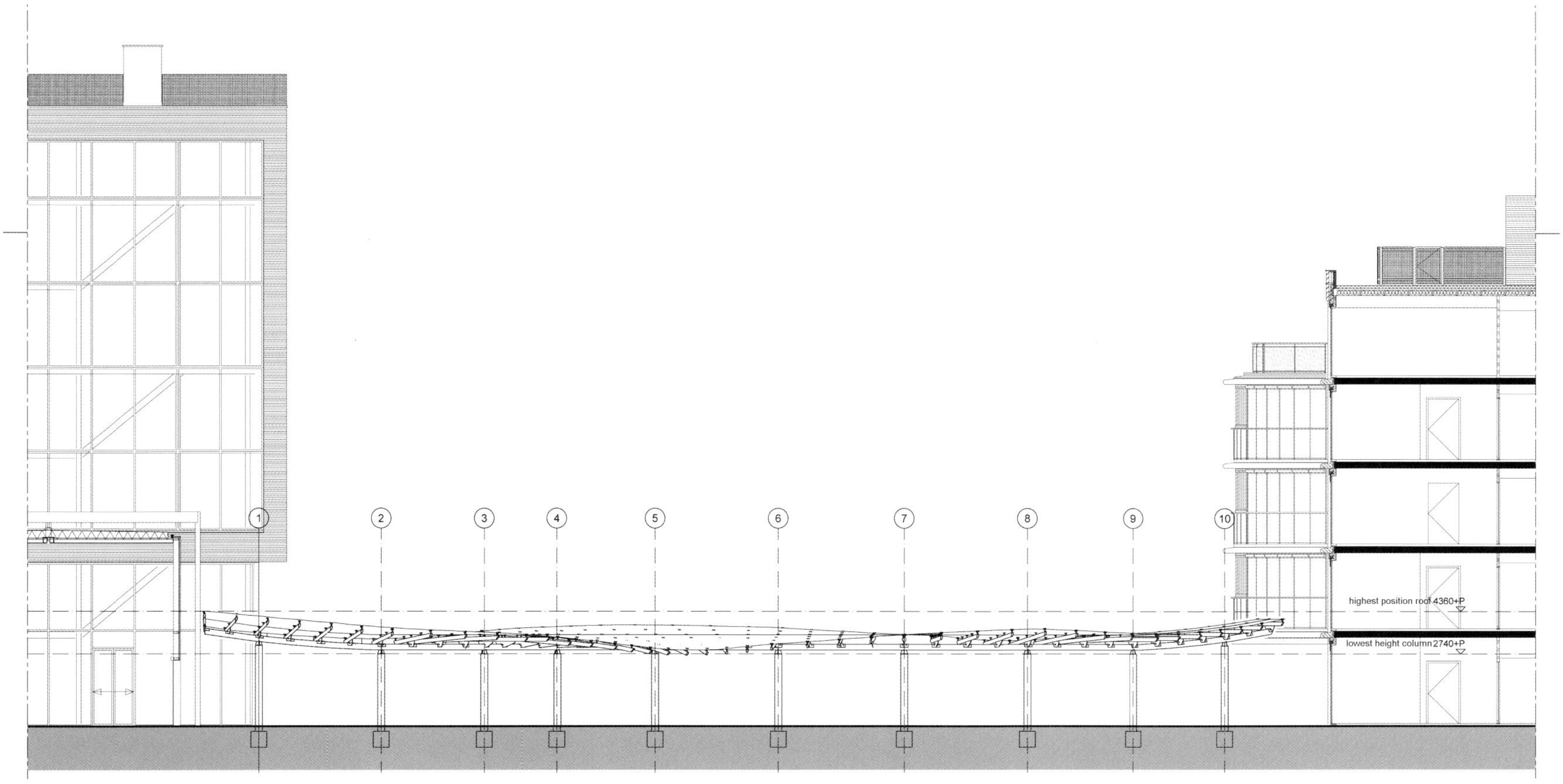

pergola drawing: Octatube and Architectuurcentrale Thijs Asselbergs

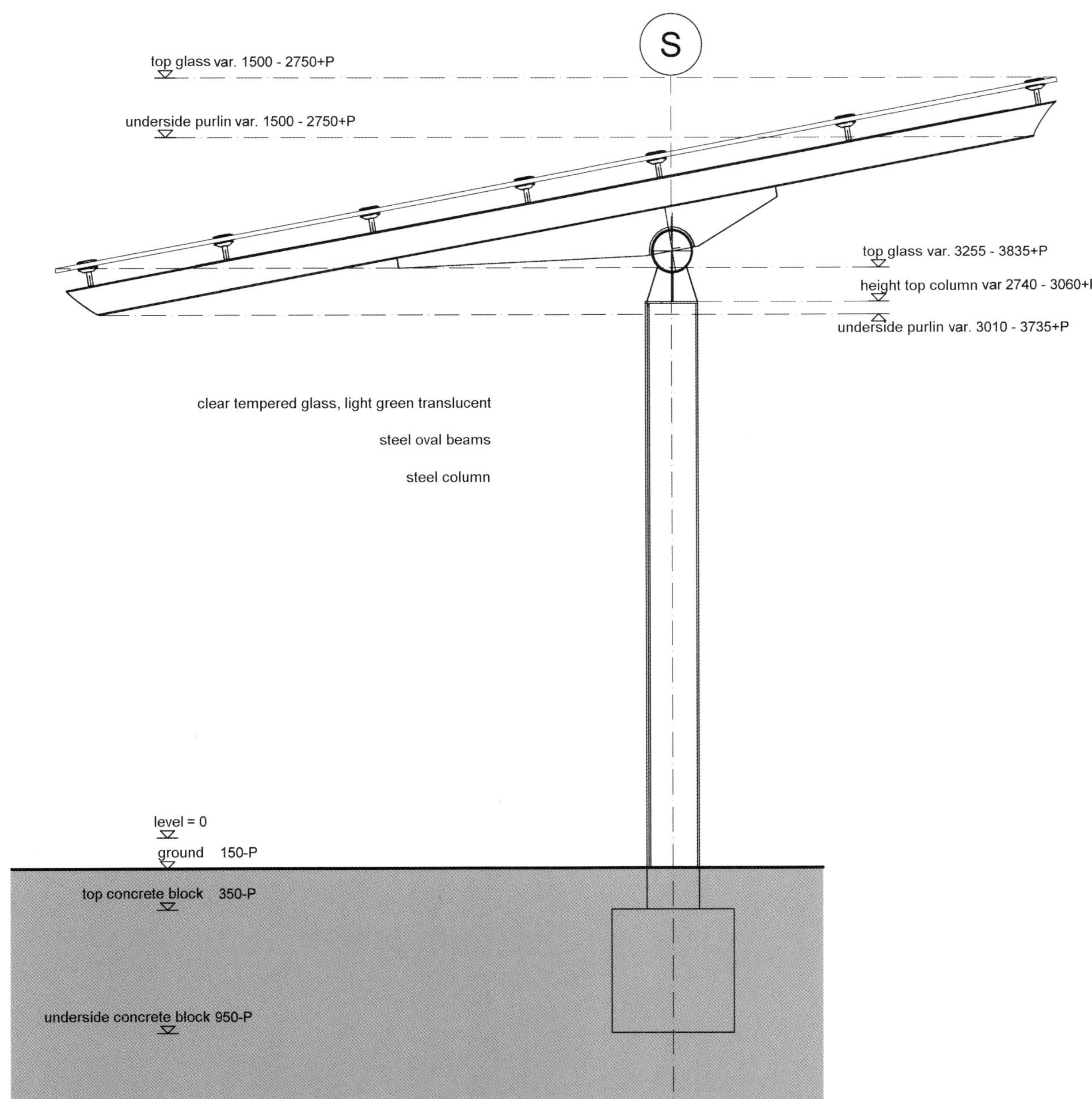

pergola drawing: Octatube and Architectuurcentrale Thijs Asselbergs

• **Landscape Architect** - Edwin Santhagens, Stephen Tas • **Firm** - Buro Sant en Co • **Area** - 5,500 m^2 • **Photographer** - Buro Sant en Co

THE HILL SQUARE

小山广场

Tilburg, The Netherlands

In this winning design, the Hill becomes the main square of Tilburg once more; lively and relaxing, modern and green.

The layout of the square is simple: a triangular shaped platform with lime trees set in a triangular space. The diverging shape of the square is defined by its surroundings and accentuated by a planting of plane trees along the edges. The triangular shape of the platform molds itself into the available space and is a reference to the rural origin of the square. The materials for the square are chosen in such a way that they easily connect with the surroundings and the new design for the city ring. The platform has lots of space and offers many opportunities for contemporary uses. With the chosen materials, the "stage" is a hybrid between an urban center and a green oasis.

The diverging shape makes it difficult to define the edges to the square. To accentuate the space and give it a strong identity, they have chosen to add a raised platform in the center of the square. By raising the plateau and giving it a coating of laminated stone, the reference to a stage is made explicit. The edges to the stage are left untreated and form a contrast to the top. The height of the platform is 40cm, which is ideal for sitting. As a consequence, the platform becomes the largest piece of urban furniture in Tilburg. The plateau is not only floating (accentuated by recessed lights), but also lively. The basalt strips are alternated with strips of grass and strips of "water". A new lime tree has been planted at the location of an existing statue that has been moved for the occasion. In due course the tree will mature and dominate the square like a historical lime tree once did.

There is a need for a central place where things are happening, a heart for the city. The Hill square has the honor of being that place.

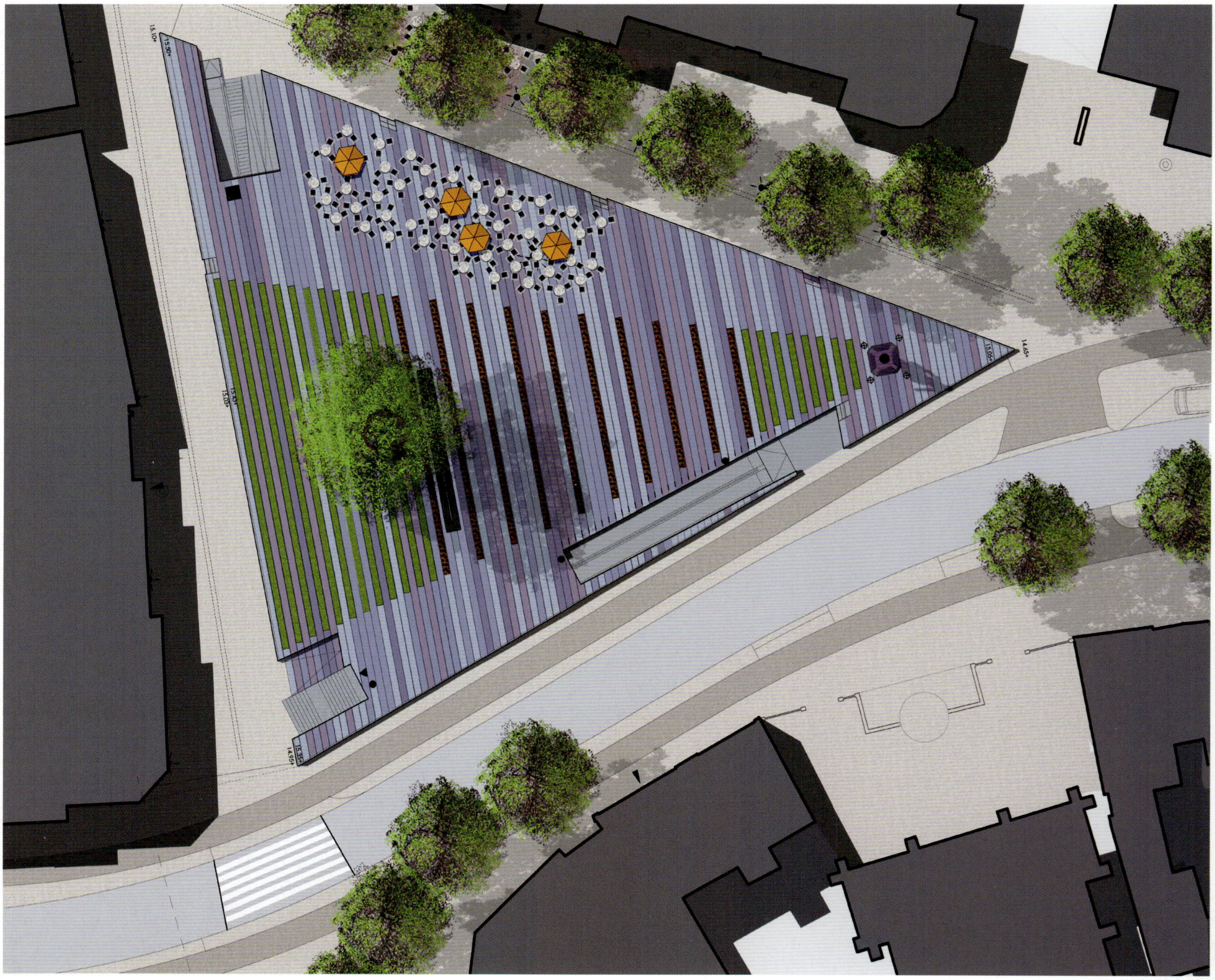

DUTCH

Philip

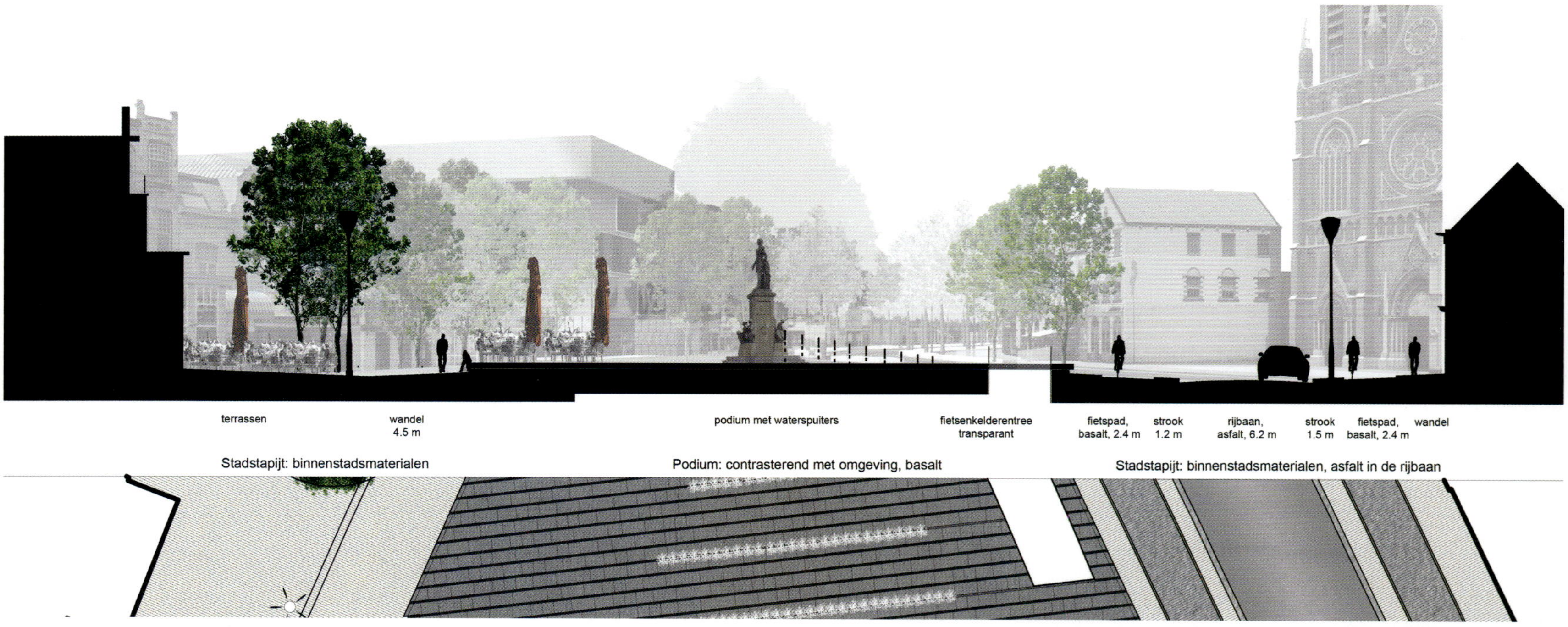

terrassen
wandel
4.5 m
podium met waterspuiters
fietsenkelderentree
transparant
fietspad,
basalt, 2.4 m
strook
1.2 m
rijbaan,
asfalt, 6.2 m
strook
1.5 m
fietspad,
basalt, 2.4 m
wandel
Stadstapijt: binnenstadsmaterialen
Podium: contrasterend met omgeving, basalt
Stadstapijt: binnenstadsmaterialen, asfalt in de rijbaan

通过这个获奖的设计方案改造的小山广场，休闲而不失活泼，现代又充满绿色，重新成为蒂尔堡的中央广场。

整个广场的外形呈三角形，让人一目了然，几个长条形的草皮点缀其中。广场周边的布局使广场显得十分发散，格外突出了种植在广场边缘的法国梧桐。三角形布局可以有效地提高广场的可用空间，同时也在一定程度上反映了广场所在地的乡村风格。广场建设的选材遵循两个原则：一是要容易融入周边的建筑风格，二是要与新的城市生活圈的规划相互协调。广场空间广阔，使用灵活，加上精心选择的建筑材料，充分体现了都市中心与绿色休闲的完美结合。

广场相对松散的外围，让人很难判断哪里才是广场的边缘。为了突出空间并赋予广场鲜明的特点，设计师们将广场中心抬高，形成一个高出地面的三角形平台。平台上面用片状石块铺设而成，简单明了，仿佛一个剧场的舞台。平台边缘不加处理，以期与平台中心形成强烈的反差。平台高 40 厘米，十分适合人们就坐。因此，这里成为了蒂尔堡最大的城市装置。在大量的隐蔽式灯光的照射下，整个平台呈现出一种飘浮的状态，同时也充满了活力。平台地面由长条形玄武岩、长条形草皮和长条形喷泉水系交错排布而成，还可以根据需要临时移动。一棵新的欧椴树种植在临时移走的现有雕像的位置。在适当的时候，种植的树木会成长壮大，就像多年前矗立在这里的那棵欧椴树一样，耸立在广场上。

城市中心地带需要欢乐的气氛，成为城市动力的来源。小山广场十分有幸，恰如其分地成为了蒂尔堡的欢乐中心。

HappyKids
happykids.nl
HappyKids
NESTEA
l'Orange

HUNDSUND COMMUNITY CENTER

哈桑德社区中心

Fornebu, Barum, Norway

Hundsund center is built around a central pedestrian zone that provides access to all the surrounding facilities. The pedestrian street has granite flooring, and a meandering stream of water runs along the center of the street, creating a refreshing element. The junior high school, the nursery school, the sports ground and the swimming pool each has its own entrance from the pedestrian street. Vehicles approach the area for loading and unloading at a roundabout at the north end of the central zone. The area is designed to become a social meeting point for the new residents of Fornebu.

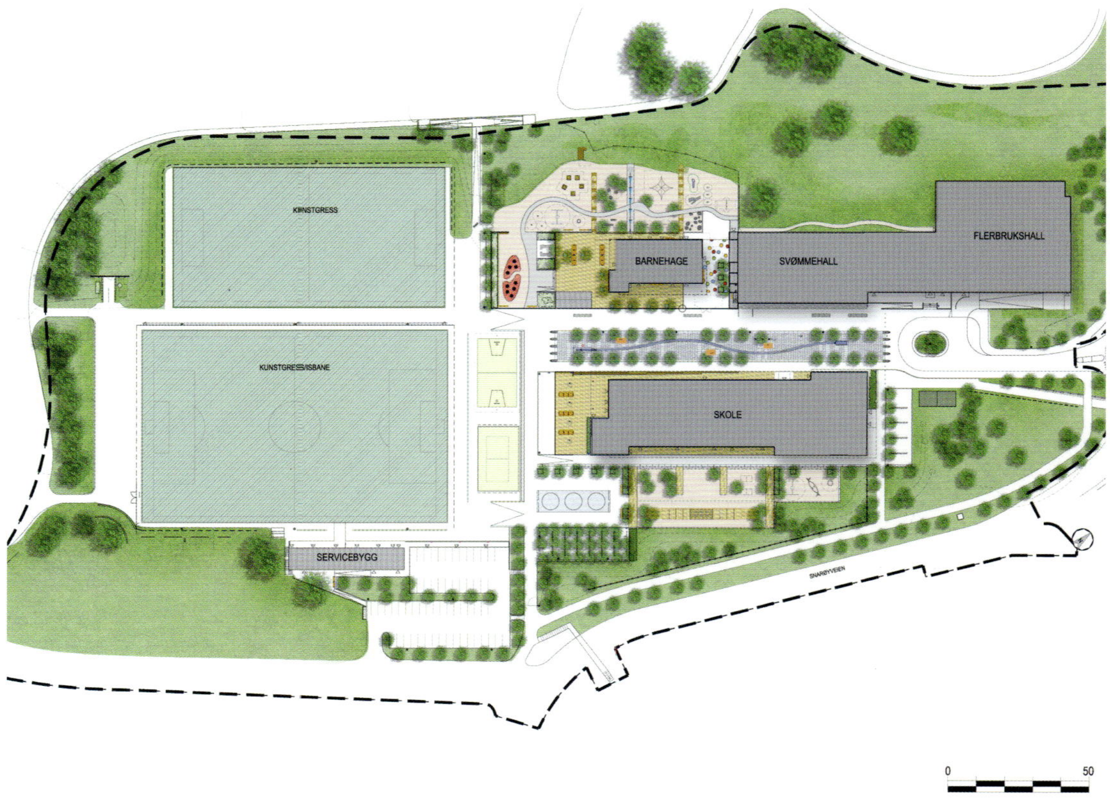

SVØMMEHALLEN

TINE

哈桑德社区中心是围绕一块步行街区建造起来的，周边遍布各种便利设施。步行小路由花岗岩板铺就而成，一条蜿蜒流淌的小河徜徉在步行街边，带来一股清新的气息。沿着步行街道，两边分布着初中校园、幼儿园、运动场和游泳池。社区中心的北部设有迂回环形路，车辆有序地通过这里进出停车场。社区中心的主要功能是为福内布地区的新居民提供一个社交和聚会的地点。

KAP 686 – THE LONG ROAD TO A NEW SQUARE

Cologne, Germany

KAP 686——通向广场的漫漫长路

The basic design concept was created by overlaying the images associated with the square, its use and its location.

People flow through the built-up urban landscape, each like a water droplet in a river. The skaters have made this flow of people into a game. Quiet, long drawn-out stretches with large radiuses alternate with jumping at obstacles, like the flowing and spraying of water in a river.

The location of the square on the Rhine places this image in an appropriate scenic context.

The implementation of this concept in reality is achieved by overlaying the area with a virtual grid which has uniform building areas at the cross-over points. The grid and building areas represent urban elements and are taken from the urban environment.

However, to turn this stark grouping on a grid into a spontaneous arrangement that is optimal for this sport, the area had to be reorganized using a particular algorithm. This turned the building areas into structures of different sizes; they rise out of the landscape or sink into it to intersperse the space with green elements, meadows and trees. The basic structures are skate objects made from concrete and stand like stones covered by water in a river of flagstones.

The ground combines with the skate objects using a template.

本项目最主要的设计理念是充分利用广场的功能和地理位置，与周边风景相互融合。

人们在这个风景如画的广场中，每个人如同河流中的一滴水，川流不息。滑板者已经将人的行走变成一种游戏，平静、冗长的大半径滑行动作与遇到障碍物时的跳跃动作间或交替，如同河流中平静的流水与跃动的水花一样。

广场紧邻莱茵河，视野开阔，风景优美。

现实中要实践这一理念则是通过在这一区域覆盖一个虚拟的网格来完成，这种虚拟网格中各个交叉点的建筑面积完全相同。网格与建筑面积象征着城市环境中的每一个元素。

然而，这种基于网格之上的刻板分组形式需要运用一些特别的运算方法，对这一区域进行重新规划，将之转变为一个更适合滑板运动的合理布局。这样的设计使实际的建筑区域面积变得大小不一，各种建筑物在景观之上拔地而起，矗立于景观之上或是掩映于景观之中，利用各种绿色元素、草坪以及树木来点缀建筑空间。广场上设有很多混凝土结构的滑板设施，如石头般矗立着，连成一条如河流般蜿蜒流淌的石板通道。

这些滑板设施通过一个模板与广场的地面相连。

Ital.

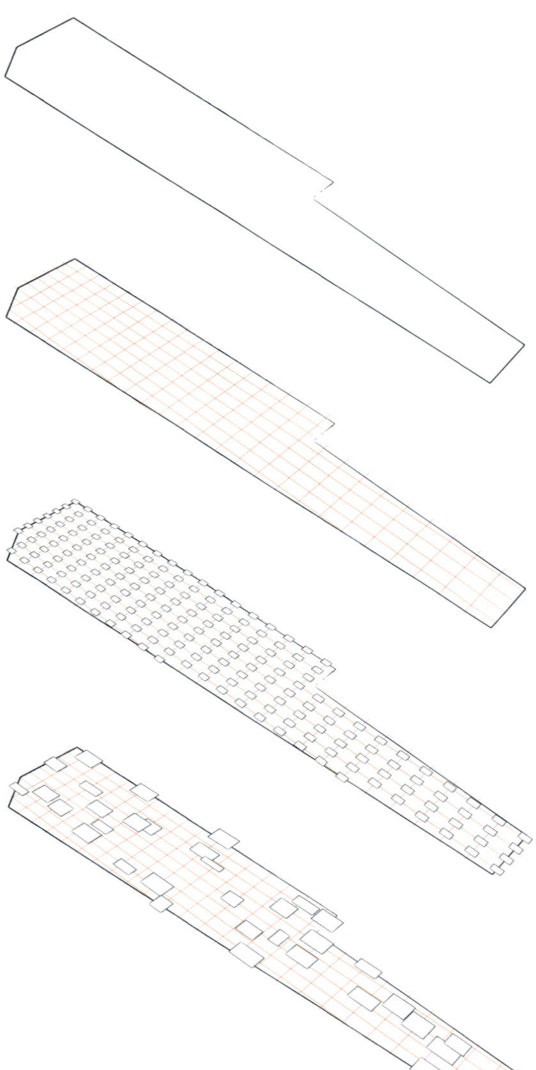

AVERBODE ABBEY SQUARE

阿夫伯德修道院广场

Scherpenheuvel-Zichem, Belgium

The square's design strives towards simplicity and homogeneity with its context. Rather than competing with the surrounding historical buildings, it only works to enhance their strong presence. Walking through the gatehouse of Averbode Abbey now, after the realization of the square's new design, one is confronted with a picturesque view of the abbey's quarters and its shimmering reflection in the water.

On the rainy days the water mirror is filled up with rainwater flowing down from the square which is collected in a slightly depressed surface creating this large puddle. The flexibility of the courtyard is guaranteed with the possibility to drain the water from its resting place, thus increasing the usable area of the courtyard for various events.

Along with the water mirror, the quality of this public space is improved by the integration of its ancient Tilia trees with old periphery wall and cobblestone path, as well as the integration of new parking spaces for visitors with reduced mobility, and sufficient facilities for those who choose to visit the abbey by bicycle. During the dark hours of the night, the square itself is dimmed by the light reflecting off the church and other surrounding buildings and LED strips are used to accentuate the direction of the pathway.

Taking into consideration the rich history of Averbode Abbey, it was imperative not to bombard its visitors with historical information, but instead to subtly introduce its history in the square itself. Hence, out of the respect for the context all of the information panels, bins, and bicycle racks have been custom designed and manufactured in dark rusted steel reminiscent of the ferrous iron sandstone appearing in the facades of the abbey buildings.

Passing by the church, one can also be reminded of the monastic rule which has been engraved in the curb stone: "one in heart and one in soul heading towards God."

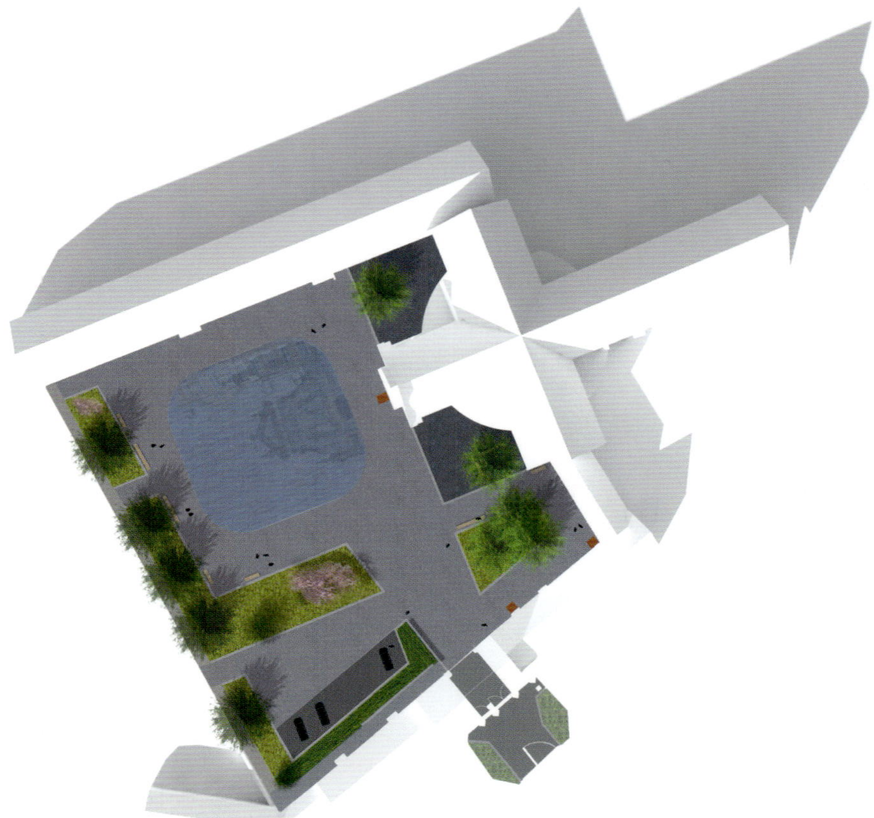

• **Area** - 5,000 m² • **Client** - abbey of Averbode

阿夫伯德修道院广场项目的设计理念是提倡一种简约质朴的风格，力求与周边环境和谐一致。广场周围历史悠久的建筑比比皆是，广场项目有效地衬托了这些建筑物的高大形象。广场建成后，游客们可以穿过阿夫伯德修道院大门，领略广场动人的设计，在波光粼粼的浅水池中欣赏修道院的水中倒影。

雨季，广场上的雨水汇集到浅水池中，避免了广场上大面积的积水。广场边上的庭院设计灵活，雨水还可以通过庭院中的楼梯向外排出，有效地提高了庭院的可利用空间。

在浅水池周围，生长多年的椴树林、古老的外墙、鹅卵石小路以及新的停车场形成了一个整体，加上充足的公共设施，有效地减少了广场上的车流量，游客们可以在这里选择骑自行车游览观光。入夜，广场上灯火通明，在修道院和周边建筑的映衬下，使用LED灯光指示的步行道路显得十分醒目。

阿夫伯德修道院历史悠久，这种历史韵味在广场上体现得淋漓尽致。因此，出于对广场的尊重，广场上的信息板、垃圾箱和自行车支架等都是根据传统设计，使用仿古的黑色铁艺制成，在后面修道院建筑那饱经风霜的铁锈色岩石的衬托下，一种怀旧的伤感油然而生。

走过修道院，游客们还可以从路边石头上的字迹中再次感受到修道士们的信念：“心灵和灵魂都属于上帝。”

• **Landscape Architect** - OLM • **Consultant** - Serero Architect / OLM Philippe Coignet and Silvia Zagheno • **Light Designer** - ON, Vincent Thiesson

SQUARE OF THE SEASONS

季节广场

Paris La Defense, France

The business district of La Defense in the North-West of Paris was built in the 1950s. Since 2010, it has started a major transformation of its public spaces around the axis leading towards the Arch. The square of the Seasons is part of this larger transformation and anticipates the nearby construction of the two giant Hermitage towers (higher than the Eiffel Tower) along the Seine River. On top of a several layers of parking lot, the project ties together diverse buildings (housing, offices and retail) and organizes the pedestrian flows through a cherry, and maple grove in keeping the existing birch trees.

The square is based on two ground materials: wood in the inner part and large concrete paving on the outskirts. All vertical ventilations emerging from the ground have been removed and hidden below the wooded platform. The elevator going down the parking lot was redesigned to be slimmer. A specific lighting offers a blue ambiance at the base of the wooded benches and vertical post distributes a warm atmosphere throughout the public place.

• **Area** - 6,500 m^2 • **Client** - EPADESA

SANDWICHS

位于巴黎西北部的拉德芳斯商业区始建于20世纪50年代。从2010年开始，沿着朝向凯旋门轴线的周边，对广场的公共空间进行了大规模的改建。季节广场是这次大规模改建项目的一部分，广场完工后，沿着塞纳河边还建造了两座高大的赫米迪奇塔（高度超过埃菲尔铁塔）。从邻近的多层停车场上方向广场望去，广场将住宅、办公楼和零售商铺等各式建筑连接在一起，在桦树、樱桃树和枫树林之间形成了一条步行区域。

广场地面由两种材料铺设而成，户外是混凝土地面，广场内部是实木地板。木质地板隔离并遮蔽了所有来自地面的垂直气流。重建后，细长的电梯直接通向停车场。在特别设计的蓝色灯光下，实木长椅和高大的立柱一起为公众提供了一个温暖舒适的活动空间。

COLONNADE COURTYARD

Berlin, Germany

科隆内德庭院项目

The vision of Friedrich Wilhelm VI is finally a reality: the old Island in the River Spree, with its important art collections is becoming a "sanctuary of art and science". After the completion of the restoration of the magnificent buildings, their open spaces will be accessible to the public for the first time. In the center of Berlin, an area is created with atmospheric places that invite you to linger and stroll, to dream and explore.

In the "Kolonnadenhof" the structure of paved areas and green spaces is developed from the ground plan of the listed courtyard. Between the stone paths, vegetation rises out of the surface. The finely developed vegetation levels give the "Kolonnadenhof" surface a spatial differentiation without placing it in the foreground. The "Kolonnadenhof" thus forms an attractive environment for the outstanding architecture of the museums and a serene area for the bronze sculptures of the "Alte Nationalgalerie".

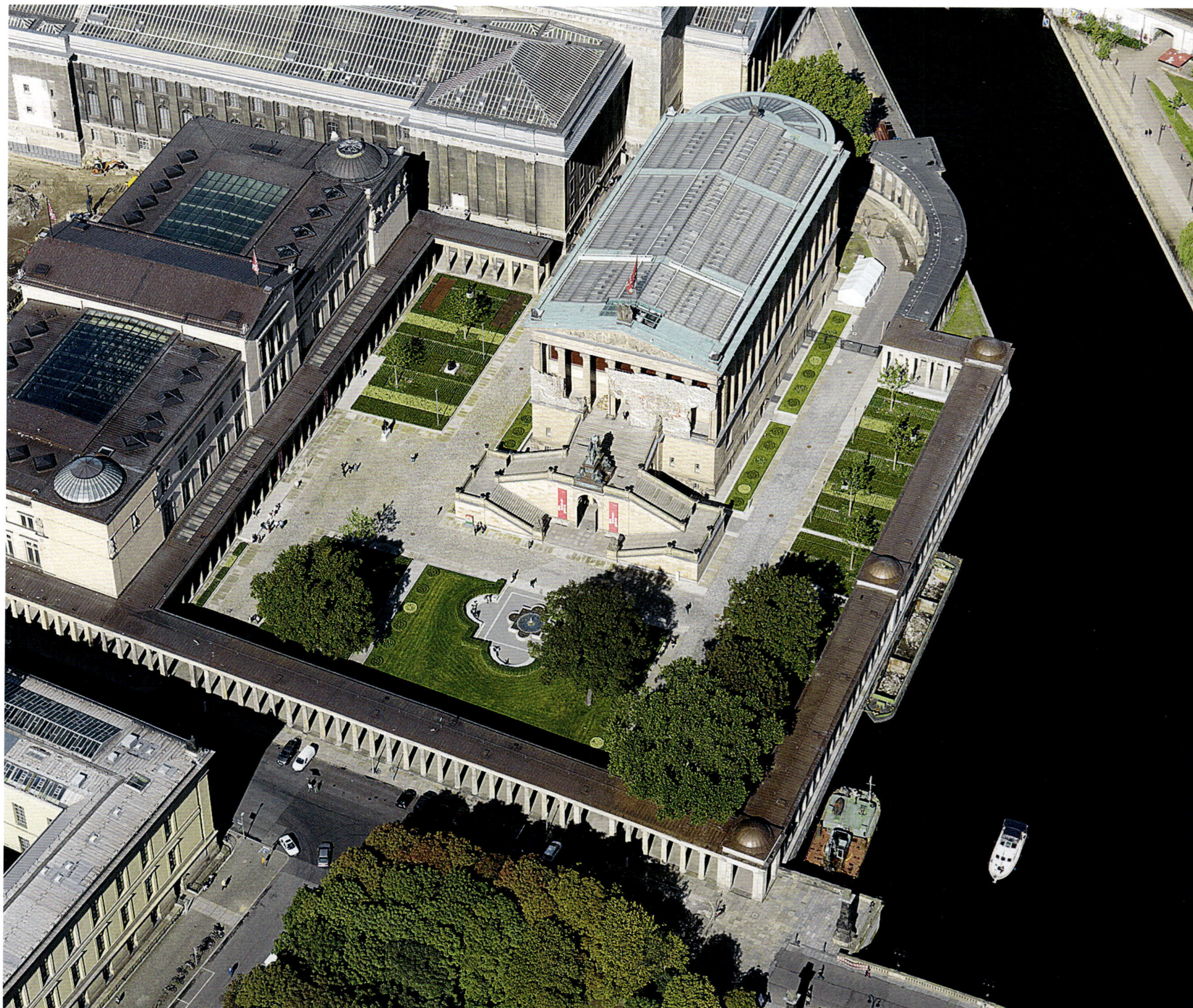

• **Photographer** - Claas Dreppenstedt, Berlin and Arial view - DOM publishers, Berlin

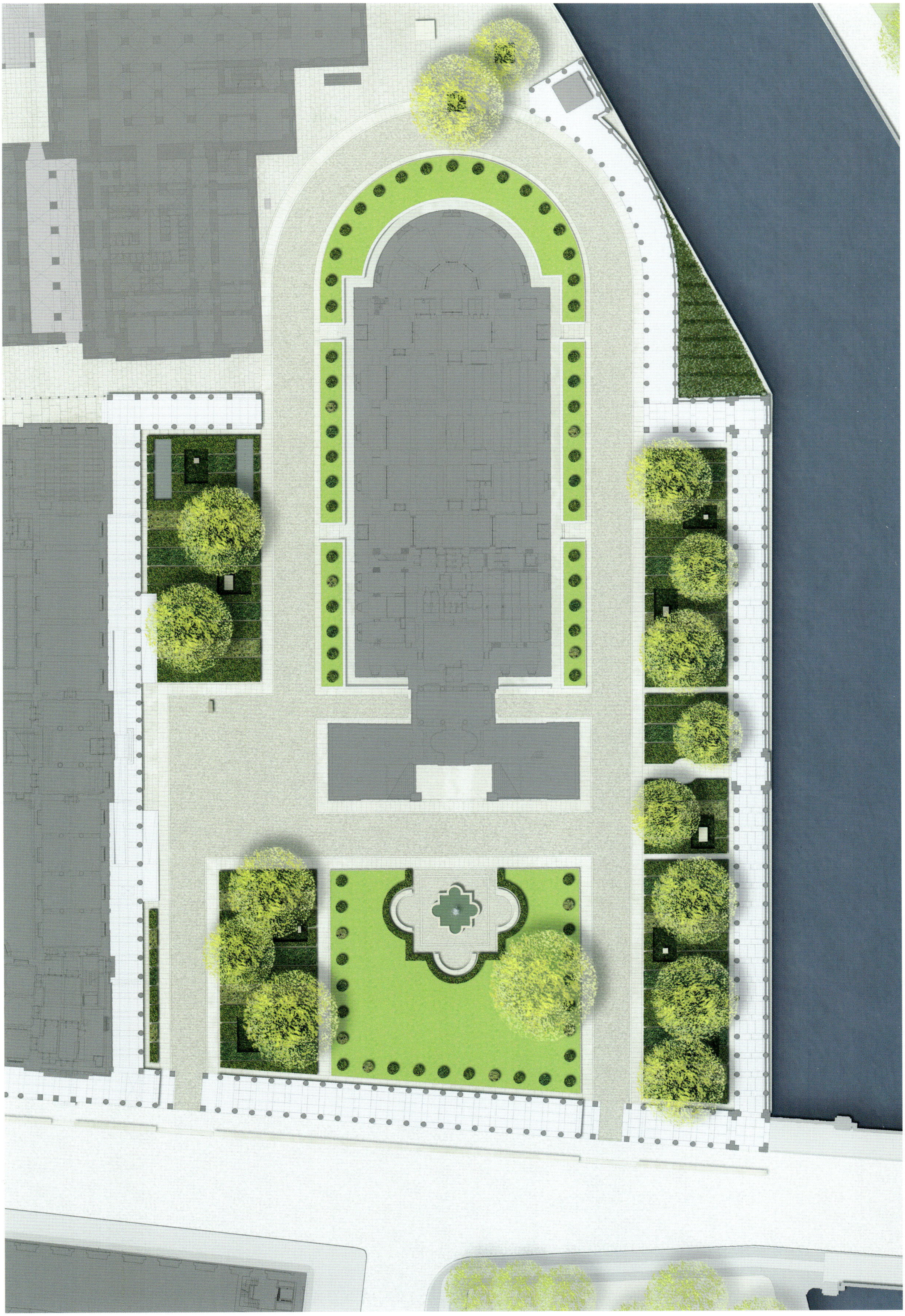

Neues Museum
Neues Museum
Alte National-galerie

科隆内德庭院项目验证了弗雷德里希·威廉四世的预言：施普雷河中古老的岛屿和岛上重要的艺术建筑，将成为艺术与科学的避难所。这些宏伟壮丽的建筑修复完工后，将在第一时间向公众开放。在柏林市的中心地带，将诞生一个高贵而华丽的地方，游客们在这里漫步徜徉，体验探寻历史的光荣与梦想。

在科隆内德庭院中，按照规划设计，铺设路面与绿地交错分布，相得益彰。石板小路两侧，绿色植物拔地而起，抬眼望去，十分舒适。精心设计的层状绿化带巧妙而清晰地将整个科隆内德庭院划分为多个区域，加上引人注目的博物馆建筑，静静矗立在老国家画廊中的青铜雕塑，将整个科隆内德庭院打造成一个具有深厚底蕴和魅力的地方。

SQUARE DORCHESTER—PLACE DU CANADA

多切斯特广场和加拿大广场

Montreal, Quebec, Canada

Named to commemorate the Confederation of 1867, Dominion Square has endured as the largest and most renowned garden square in Montreal. The southern and northern portions were respectively renamed Place du Canada in 1966 and Dorchester Square in 1987. Dorchester Square was Canada's wealthiest neighborhood at the end of the nineteenth century. Built between 1876 and 1880 upon a desacralized cemetery where 50,000 human remains are still buried, the square is significant in terms of archeology, architecture, and landscape history.

Despite the important presence that Dominion Square has played throughout Montreal's history, it nonetheless was subjected over time to alterations and urban pressures that ultimately diminished the quality of the park experience. However, it is interesting to note that the park layout and the relationship of the square to the surrounding urban fabric have remained intact, which, along with the regularities of the surrounding streets, have contributed to the preservation of an exceptional ensemble and balance between architecture, street and landscape. The alterations that did occur resulted in a transformation of the square's geometry, a reduction space for public use, and a suppression of important visual links. Poor management of the site environment also degraded the park from its original form and quality.

The accumulation of these alterations over the years preceding restoration were manifested in the loss of character throughout the green parterre, an incoherence in the distribution of tree species, pedestrian pathways that deviated from the original circulation, and an absence of urban furniture.

Although distinctively contemporary in its design approach and methodology, the restoration of Dominion Square needed to uphold its original identity and sense of place. A simple and meaningful approach was taken to preserve and enhance the integrity of the site's archeological heritage, sustainability of plant species, as well as spatial qualities and the overall landscape experience. The rejuvenation of the park's comfort and splendour has been achieved by restoring the parterre motif of lawn, reconnecting the site to its context, increasing the surface area of the Square, as well as reconfiguring and highlighting entrances (particularly through the addition of water features).

This project aims to restore the original Victorian public square, enhance links to urban generators of social activity, and remove incompatible uses to this classic oasis of grandeur in one of the city's most venerable districts.

• **Photographer** - Consortium Cardinal Hardy, Claude Cormier + Associés inc, Denis Tremblay, Marc Cramer

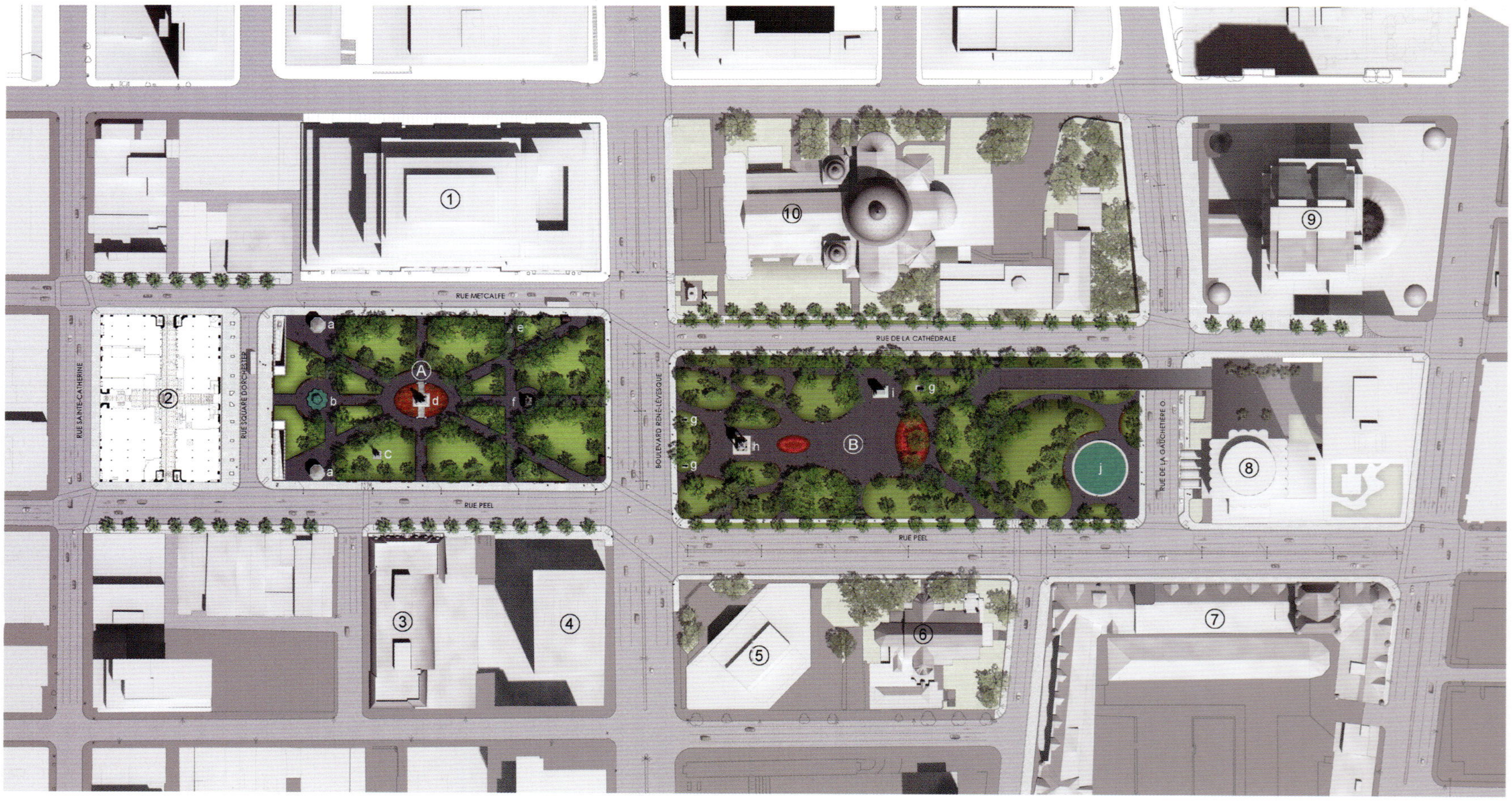

SQUARE DORCHESTER / PLACE DU CANADA

A. Square Dorchester

B. Place du Canada

BUILDINGS

1. Sun Life Building
2. Dominion Square Building
3. Le Windsor
4. Tour CIBC
5. La Laurentienne
6. St. George's Anglican Church
7. Windsor Station
8. Marriott Château Champlain Hotel
9. 1000 de la Gauchetière
10. Mary Queen of the World Cathedral

MONUMENTS

a. *Vespasienne*

b. Fountain

c. Robert Burns

d. Boer War

e. The Lion of Belfort

f. Sir Wilfrid Laurier

g. Cannon

h. John A. Macdonald

i. Cenotaph

j. Basin

k. Mgr. Bourget

CATHÉDRALE MARIE-REINE-DU-MONDE
B
RUE DE LA CATHÉDRALE
CÉNOTAPHE
RUE PEEL
LA LAURENTIENNE

自治领广场因纪念 1867 年加拿大联邦成立而得名，一直是蒙特利尔面积最大、最著名的广场。广场的南部区域于 1966 年更名为加拿大广场，广场的北部区域于 1987 年更名为多切斯特广场。其中，多切斯特广场曾是加拿大最富裕的街区。自治领广场建于 1876 至 1880 年间，所属地块曾为一处公墓，至今仍有 50000 具人类遗骸。该广场无论是对考古学、建筑学还是园林历史，都具有十分重要的意义。

尽管自治领广场在蒙特利尔的历史上占据重要地位，但随着时光的流逝，原有的广场设施和条件已经无法满足城市发展的需求。值得关注的是，改建项目没有对广场布局和周边建筑物进行过多地变动，并且顺应周边街道的建筑风格，在所保留的建筑、街道和广场园林布景之间进行了精心的权衡。历史和时间改变了广场的整体形状，减少了一些公共空间，对部分景观视野进行了压缩处理。原有的广场环境管理体系已经过时，对广场的形象和质量造成了一定的负面影响。多年来，原来的广场在旧的管理模式下所积累的大量问题不断暴露，包括绿化、树种分布、人行步道等，另外，还缺乏城市化所必需的附属设施。

尽管设计师们可以运用当代先进的方法对自治领广场进行改建和整修，但是设计方案还需要注意对原有的特点和风格进行保持和维护。设计师们选择了一种简单有效的方法，完好地保留了原有的拱门等特色建筑，同时增加了植物物种，提高了空间质量和整体园林景观效果。广场的改建工程为市民带来了更舒适、更壮观的感受，恢复了大片草地，重新设计规划了广场与周边区域的连接方式，增大了广场面积，重新规划并利用人工水系在视觉上突出了广场大门的形象。

广场改建工程致力于在保留原有广场维多利亚风格的基础上，增强广场与周边区域的联系，营造更多的社交空间，拆除不必要的设施，让这个极具历史意义的广场在城市中最庄严的地区重新焕发应有的风貌。

GÉNÉRAL DE GAULLE SQUARE AND SAINT VAAST SQUARE

高卢将军广场和圣瓦斯特广场

Armentieres, France

"The great question of town-planning is no longer the one that preoccupied Alberti, to know how to choose a site where the town should be built, but how we will manage to inherit, through a few projects, sites which are from now on all concerned by urban transformation of territories."

Here in Armentieres, this question was for the designers' essential because of the turbulent history caused by stratifications and brutal juxtaposition; after the destruction of the city centre during the Second World War, it was obvious that this main "accident" and the deep modification of the urban space were to be explained by the project. The three main monuments, the town hall, the theatre and the church were almost "lost" in the vacancy. In the late twentieth century the modification had hidden the river Lys and the geographical memory was almost lost. Thus, the project was to be settled on these two fundamental points: The site as foundation and its usages as regulation ideas.

From now on the designers could reorganize the public squares on both sides of the town hall, one mineral along the main historical road ready to accept the weekly farmers market, the other one, in a lawn at the backyard, green link to the church and opened to urban games as a kind of "boulingrin". Around the church and the theatre, the spaces are linked in a dialog with each monument by brick enclosures. The transversality and the shaping through three wide ramps are directed toward the old Lys and allow building a relationship between the squares and the future greenway.

This work by the answer brought to urban modern life with its shops, café terraces, playground…can be considered as a palimpsest way to rewrite urban centrality.

ARMENTIÈRE - L'EGLISE - LE VIVAT ET L'ACCÉ A LA PROMENADE DE LA LYS

“关于城市规划最大的问题已经不再只是关注城市本身，如怎样选择合适的地点建造城市，而是我们如何通过整个城市范畴内的几个项目或几个地点的建设和关注，设法将城市的风格和内涵继承下去并发扬光大。”

在阿尔芒蒂耶尔市，由于历史上残酷的兼并导致的一段混乱的历史状态，使得城市规划问题完全取决于建筑设计师们；第二次世界大战期间，整个城市中心被摧毁，在那之后，人们开始意识到城市的摧毁和重建工作的意义。三个主要的纪念碑、市政厅剧院和教堂在那个时期几乎消失。20 世纪后期，城市的重建工程完工后，利斯河几乎埋没在钢筋水泥的城市之中，很多标志性的地理记忆也随之消失了。因此，本项目的主旨在于解决两个最基本的问题：项目选址和项目用途。

从现在开始，设计师可以在市政厅的两侧重新规划广场的设计，一侧毗邻主要的历史街道，这里已经成为每周开放的农贸市场所在地；另一侧是后院的绿色草地，直接通向教堂，并对市民开放，人们可以在这里游戏或比赛。教堂和剧院的空间相互连通，旁边是由砖砌成的三个纪念碑。纪念碑外形呈斜坡状，上端巨大的横断面指向古老的利斯河道，矗立在广场和未来规划的绿色通道之间。

本项目最终以现代城市生活的方式，利用商店、露天式咖啡屋、操场等功能空间，重新规划了城市中心，是城市重建工程的典范。

N
0 100 500m

N
0 50 200m

VIVAT
Passage du VIVAT sur l'ancienne Lys
A Armentière

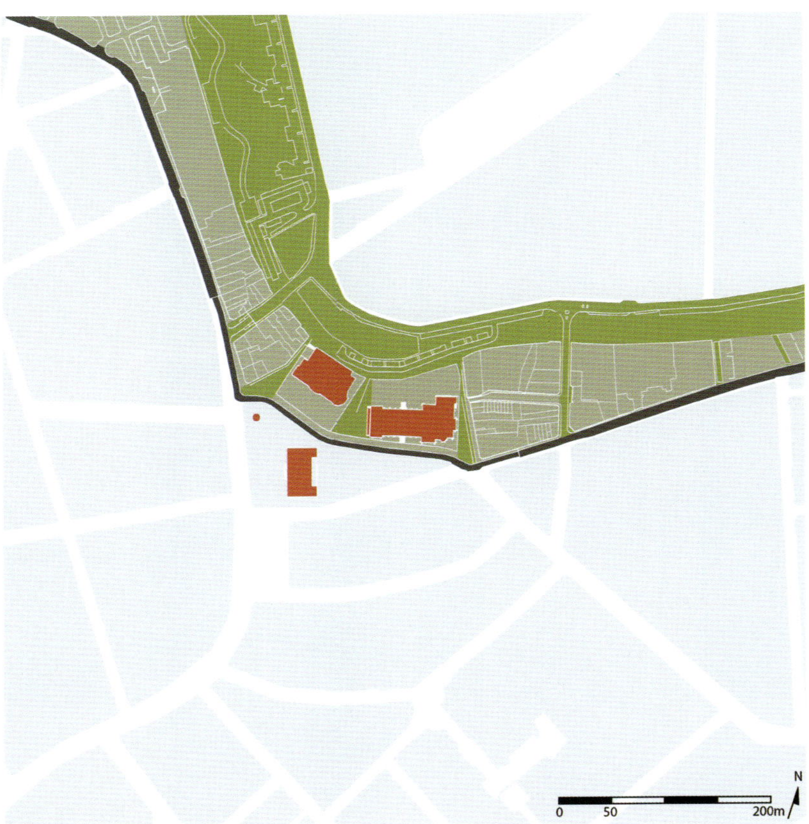
N
0 50 200m

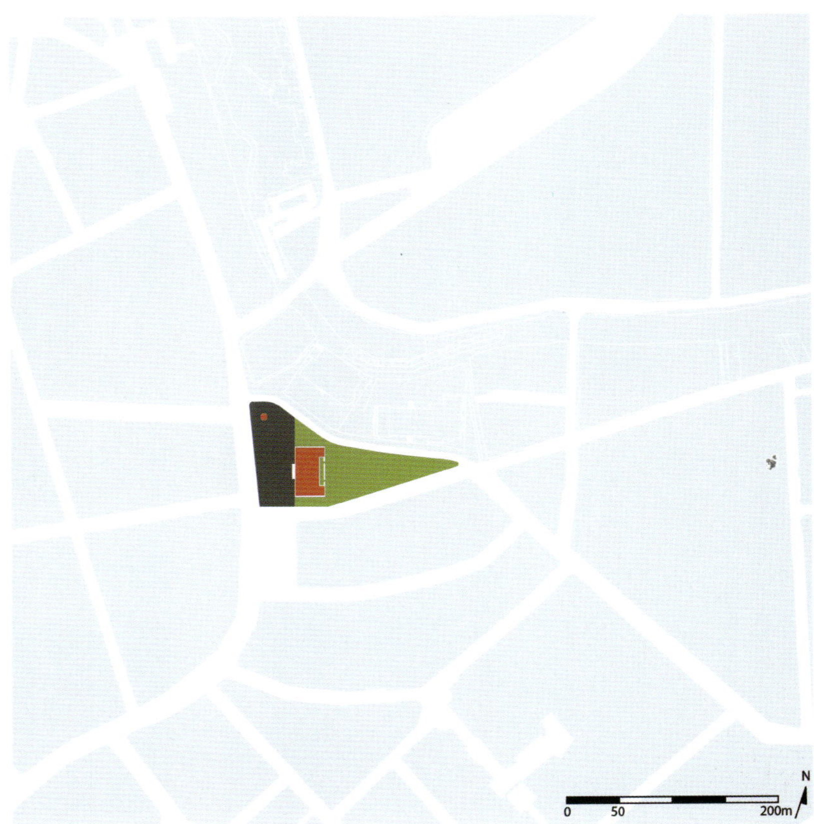
N
0 50 200m

EVASION

TIME CHIPS
SANDWICHERIE

Armentière
Parvis de l'église

PAMPULHA SQUARE

帕姆普拉广场

Belo Horizonte, Minas Gerais, Brazil

The Square was built at the last large public space in the surroundings of Pampulha, close to St. Francis Church, one of Niemeyer buildings around the lake. The project redesigned topography creating slightly sloping green surfaces that improve natural drainage, defined permanence areas – grades, benches, platforms – and cover facilities – public toilets and bar – improving the open public character of the square. Destined to large-scale events, the design defined a large dry esplanade closer to the lake, while increasing the green presence in the other direction, defining otherwise a more bucolic intimate scale for the neighborhood use. Vertical totems for lighting liberate the esplanade. The huge area associated to a low budget oriented a strict choice of low cost and resistant materials: the traditional brick in pavements and vertical elements; concrete for totems; crushed stone and grass for the permeable surfaces.

Materials:

Floor: bricks, red coloured concrete blocks, gray coloured concrete blocks, crushed bricks and grass.

Ilumination posts: concrete and ceramic tiles (artistic panel)

Trees: Tibouchina granulosa, Roystonea oleracea, Caesalpinia ferrea, Calycophyllum spruceanum, Cariniana legalis

Other: stucco painted orange and yellow, metallic handrails

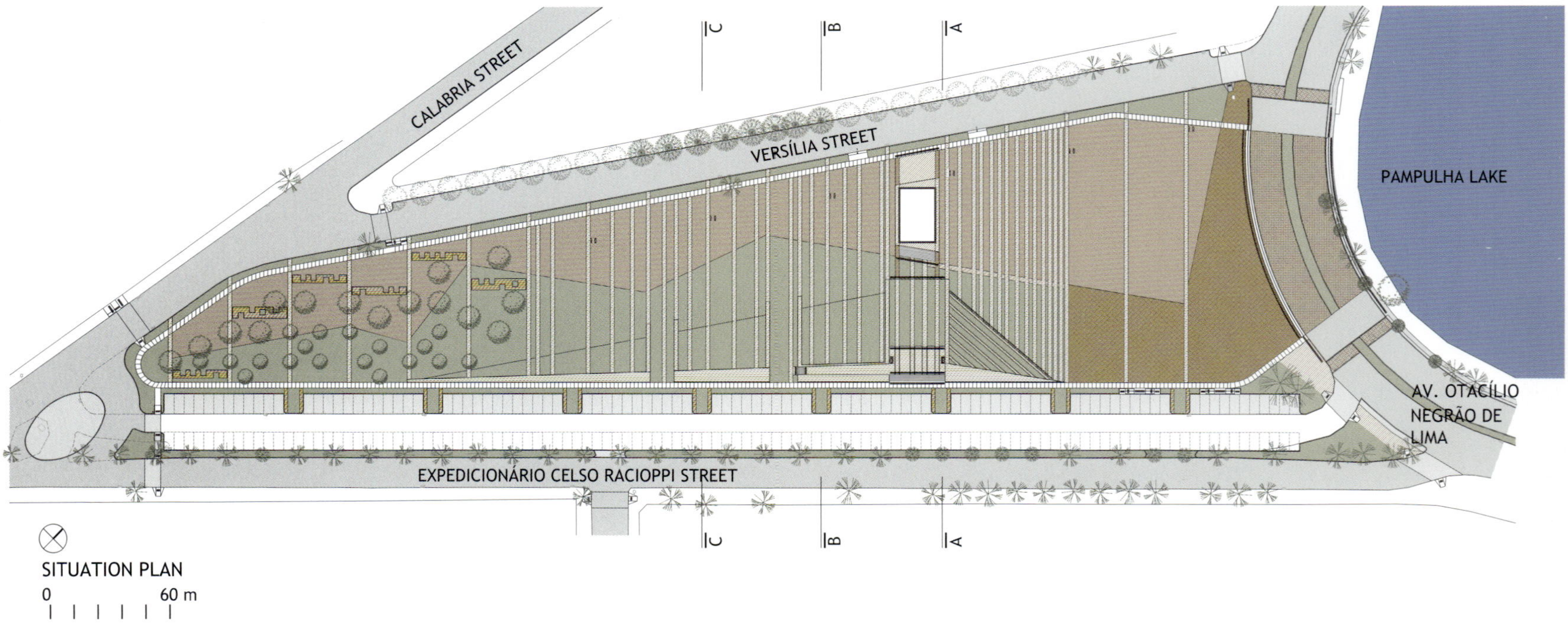

• **Collaborator** - Enara Paiva, Michelle Andrade, Henrique Boabaid

• **Firm** - Arquitetos Associados

• **Area** - 19,000 m^2

• **Photographer** - Eduardo Eckenfels

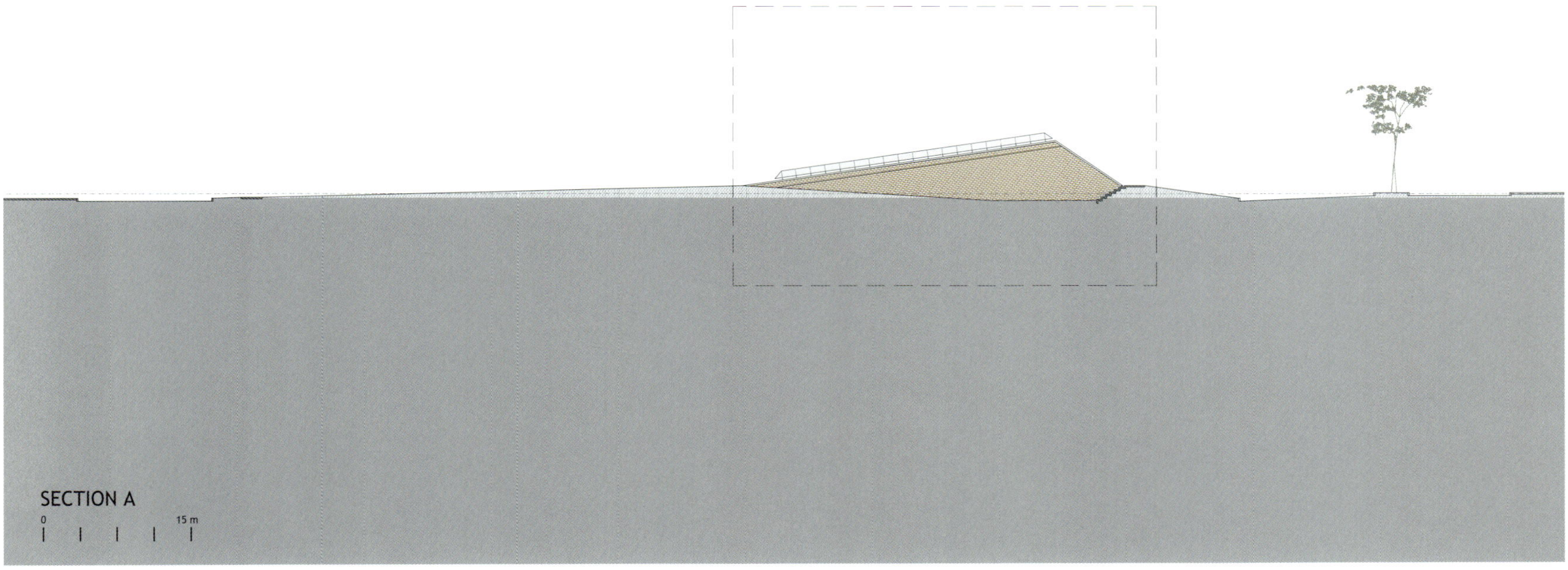
SECTION A
0
15 m

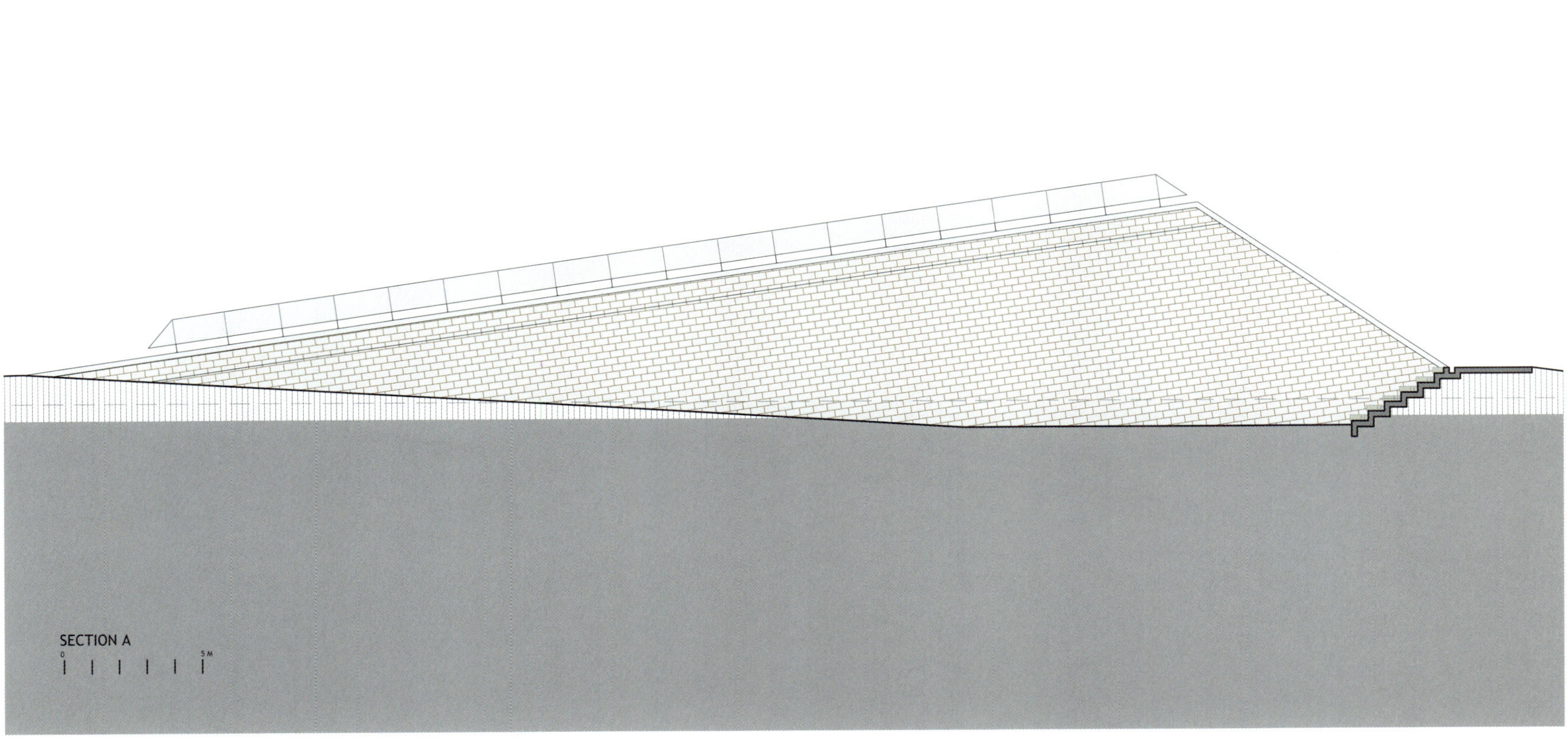
SECTION A
0
5 M

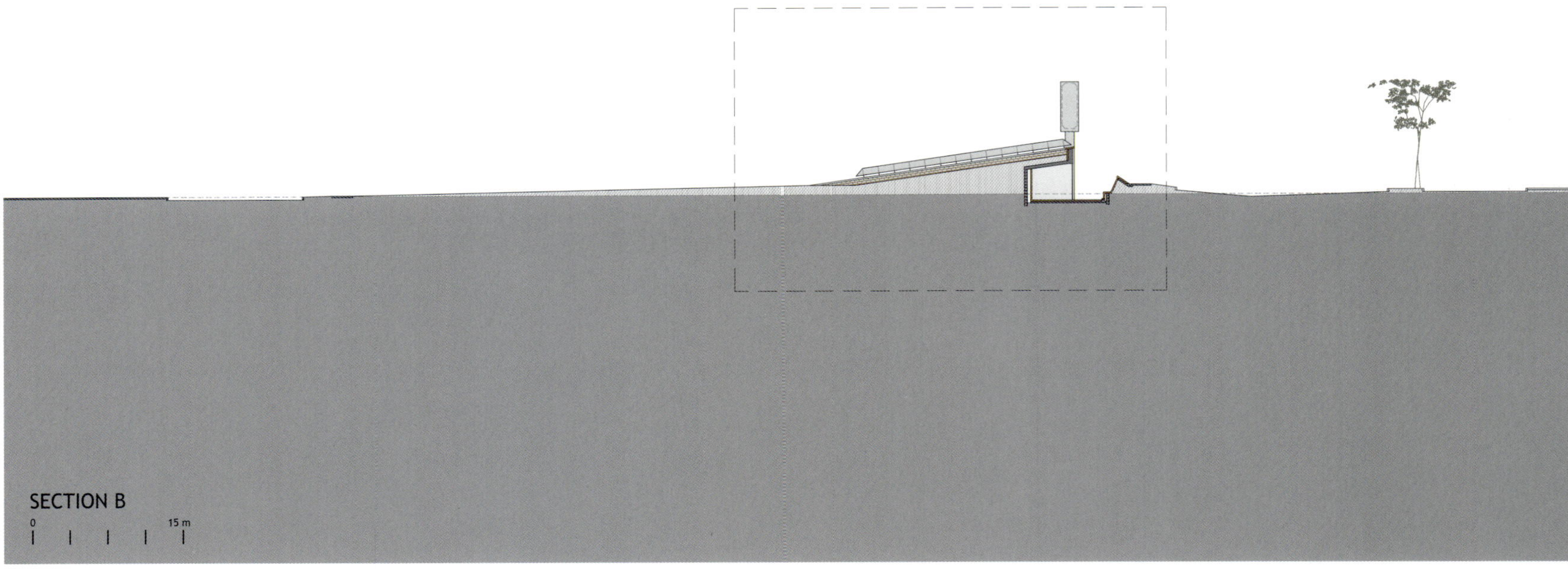
SECTION B
0
15 m

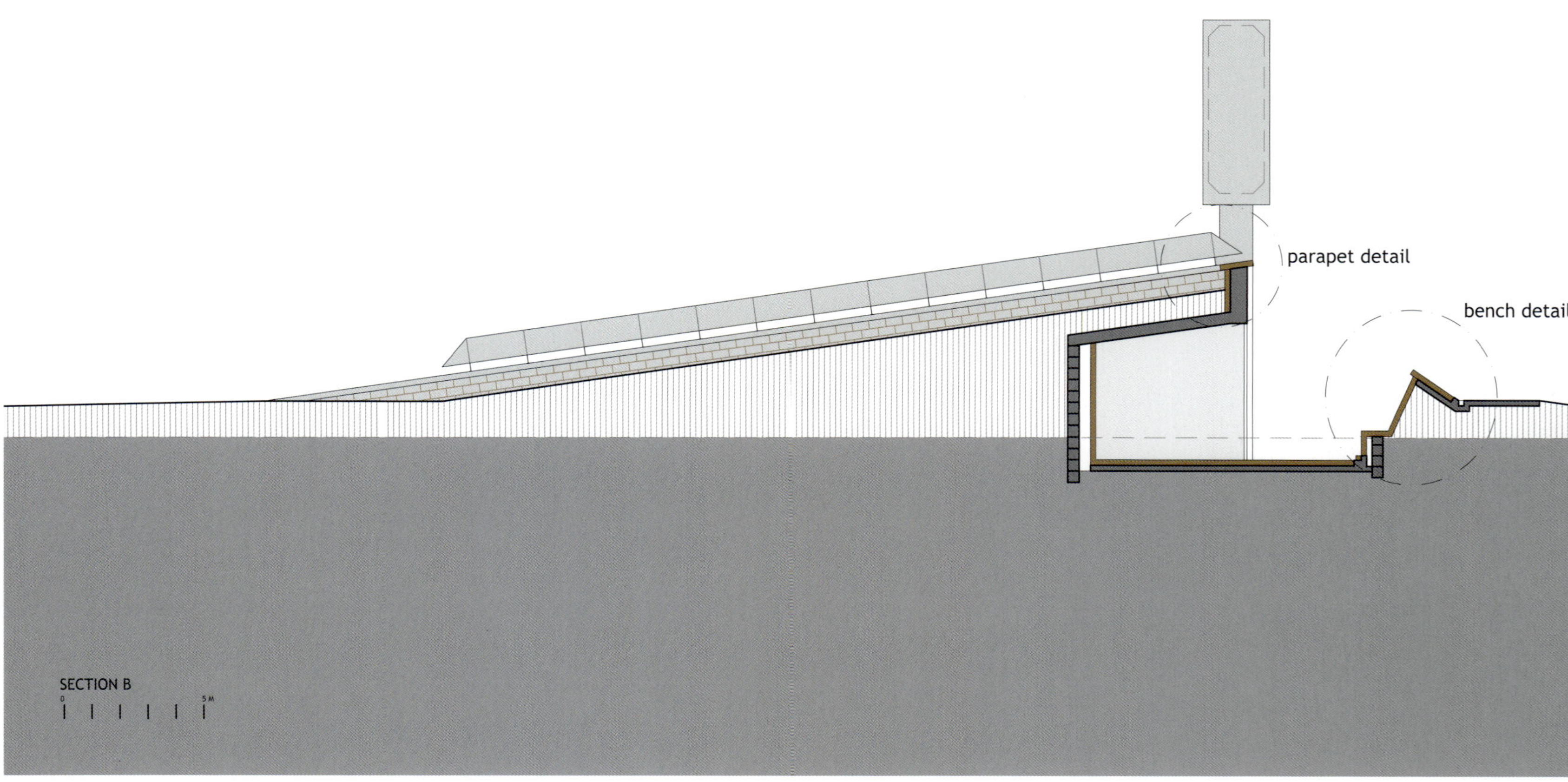
parapet detail
bench detail
SECTION B

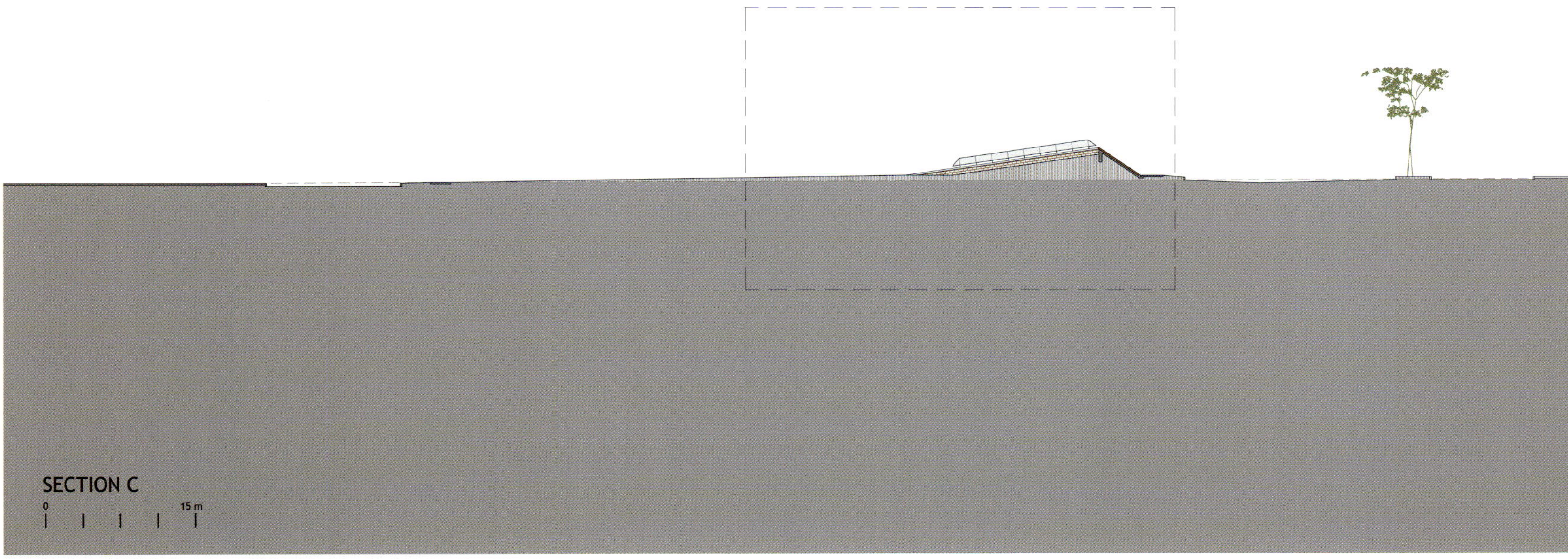
SECTION C
0
15 m

SECTION C
0
5 M

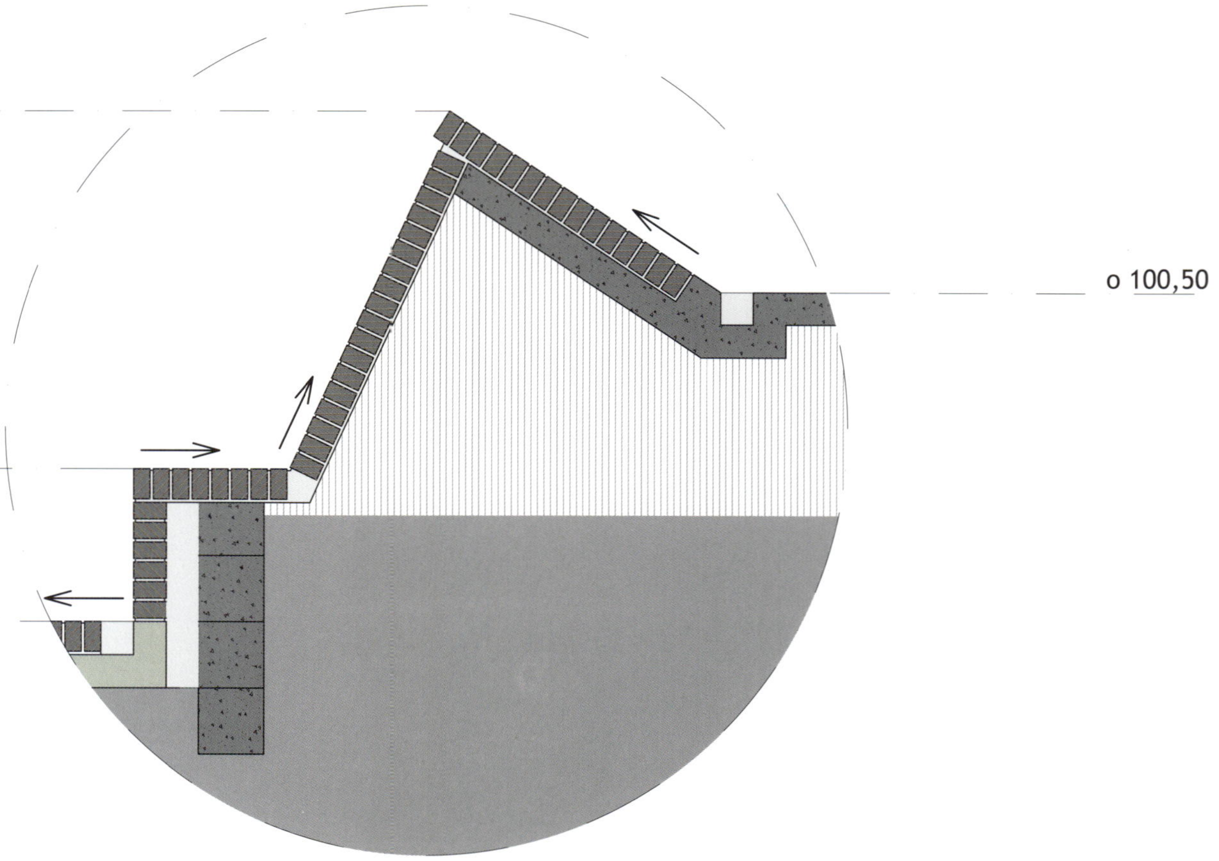

BENCH DETAIL

帕姆普拉广场拥有帕姆普拉附近地区最后一个大的公共活动空间，毗邻位于湖边的尼迈耶式建筑——圣弗朗西斯教堂。设计师们对地形进行了重新规划，设计了略带坡度的绿地，提高了自然排水能力，并设计了各种持久性设施（台阶、长椅、平台），建造了一些封闭式建筑（公共卫生间、酒吧），大大提升了公共空间的服务品质。为了给一些大型活动提供场地，设计师们在湖边设计规划了大片相对干燥的空地，在其他方位规划了大片绿地，体现了一种极富私密性的田园风情。在垂直竖立的高大图腾的映衬下，平坦的步行空间显得格外空旷。如此大面积的工程预算很低，需要通过严格的低成本筛选和选择耐用材料来完成：利用传统砖砌方式铺设步行通道；混凝土浇筑图腾；利用碎石块并种植草皮装饰兼具渗透性功能的地面。

建筑材料

地面：砖、红色混凝土模块、灰色混凝土模块、碎砖块、草皮

墙体装饰：混凝土、陶瓷砖（艺术嵌板）

树种：角茎野牡丹、菜王椰、巴西铁木、云杉光皮树、全缘叶卡林玉蕊木

其他：粉质灰泥（橙色、黄色）、金属扶栏

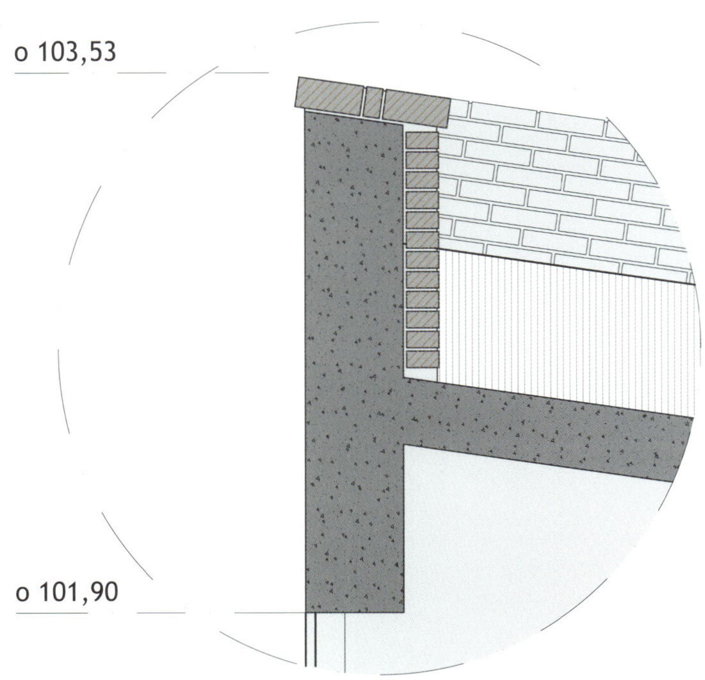

PARAPET DETAIL

• **Landscape Architect** - Østengen & Bergo AS • **Area** - 2,500 m^2 • **Photographer** - Østengen & Bergo AS

SCHANDORFF SQUARE

Oslo, Norway

斯堪多夫广场

The square is surrounded by some of Oslo's well-known motives: The Trinity Church in the south, and the Deichmann Library's pompous pillars in the east. The declining street Keysersgate opens a sight line from west.

The square creates an important slope of 7 meters height difference between two streets, making it a challenge for accessibility design to cross the square without steps. This has resulted in a curving walkway which crosses from the lower to the higher street with a slope of maximum 6.5 %, with regular resting plateaus. The walkway has become an essential and dominating feature of the square, and is therefore given a sculptural design that characterizes the area positively. Promenading, moving from lower to higher have become an important way of experiencing the green space. Along the ramp small plazas and vegetated surfaces meet in a finger-woven pattern. The old existing sculpture of the Apostle of Greenland, Hans Egede, has finally got rid of surrounding parked cars and can now rest calmly in a decent green setting.

The intended design has been to create a central urban green space with simple but sober means. The walkway's body – surface and retaining walls – are made of concrete cast in situ. The small plazas are covered with light grey granite paving, resting on walls performed of Corten steel, creating an overall warm tone.

The lawn contributes to a peaceful neighborship to the Trinity Church. Two large trees are carefully preserved through the construction period, and together with mass plantings of flowering iris, catnip, lavender and lady's mantle to create a lush and beautiful effect, mainly in blue through the season. A planted cherry grove with white spring blooming will soon contribute to a healthy and good urban environment in a dense area.

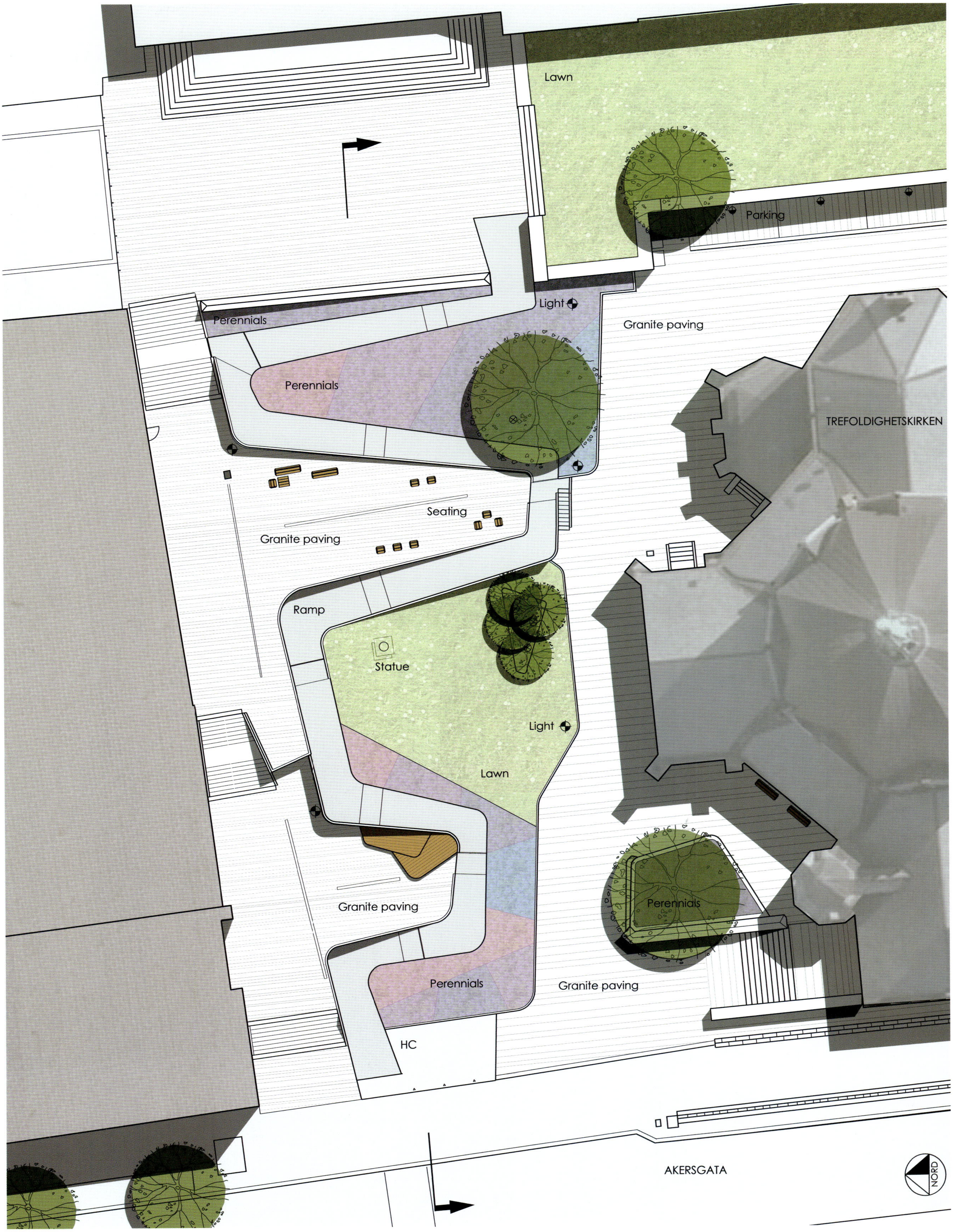
Lawn
Parking
Light
Perennials
Granite paving
Perennials
TREFOLDIGHETSKIRKEN
Seating
Granite paving
Ramp
Statue
Light
Lawn
Granite paving
Perennials
Perennials
Granite paving
HC
AKERSGATA
NORD

FREDENSBORGVEIEN
lime tree
new cherry trees
lime tree
AKERSGATA
perennials
ramp
ramp
ramp
ramp
ramp
perennials
HC Parking
29.00
28.00
27.00
26.00
25.00
24.00
23.00
22.00
21.00
20.00
19.00

斯堪多夫广场周围都是奥斯陆知名的建筑物：南面是著名的三一教堂，东面是德士门图书馆巨大的浮雕石柱，西面是呈下降趋势的笔直的凯瑟佳特大街。

广场两侧街道的水平落差接近 7 米，由于整个广场不设任何台阶，对于设计规划来说极具挑战性。设计师最终巧妙地设计了一条弯曲的步行通道，由低至高贯穿整个广场，最大坡度仅 6.5%，沿通道有规律地设置了供休息的平台。步行通道是广场的主要设施，在路边的绿地上放置一些雕塑，为整个广场增添了一丝生气。人们可以顺着步行通道，由低向高穿过整个广场，享受和体验这个绿色空间带来的舒适。小型的活动空间和石板地面呈锯齿状分布在步行道边，往日喧嚣的车流已不复存在，原有的格陵兰岛传教士汉斯·埃格德的雕塑肃穆地矗立在路旁的绿地中，仿佛在享受广场上的安逸和静谧。

设计师们致力于利用简约的风格和朴素的方式，为市中心地区打造一个绿色空间。步行通道地面和侧面的护壁都由混凝土浇筑而成。小型活动空间的地面上铺设了浅灰色的花岗岩石板，侧面使用耐候钢形成护栏，营造出一种温馨的氛围。

广场旁边的草坪为邻近的三一教堂提供了一丝闹市区难得的幽静。建筑师们在施工过程中十分小心谨慎，完好地保持了两棵古树的原貌，并在周围种植了大量的鸢尾草、猫薄荷、薰衣草和斗篷草等，春夏两季，满眼都是郁郁葱葱，绿意盎然。另外，还有一个樱桃园，开花季节，满树雪白的花团，与周边稠密的市区环境形成强烈的对比。

SQUARES IN PUIGCERDÀ

Spain

普格瑟达广场

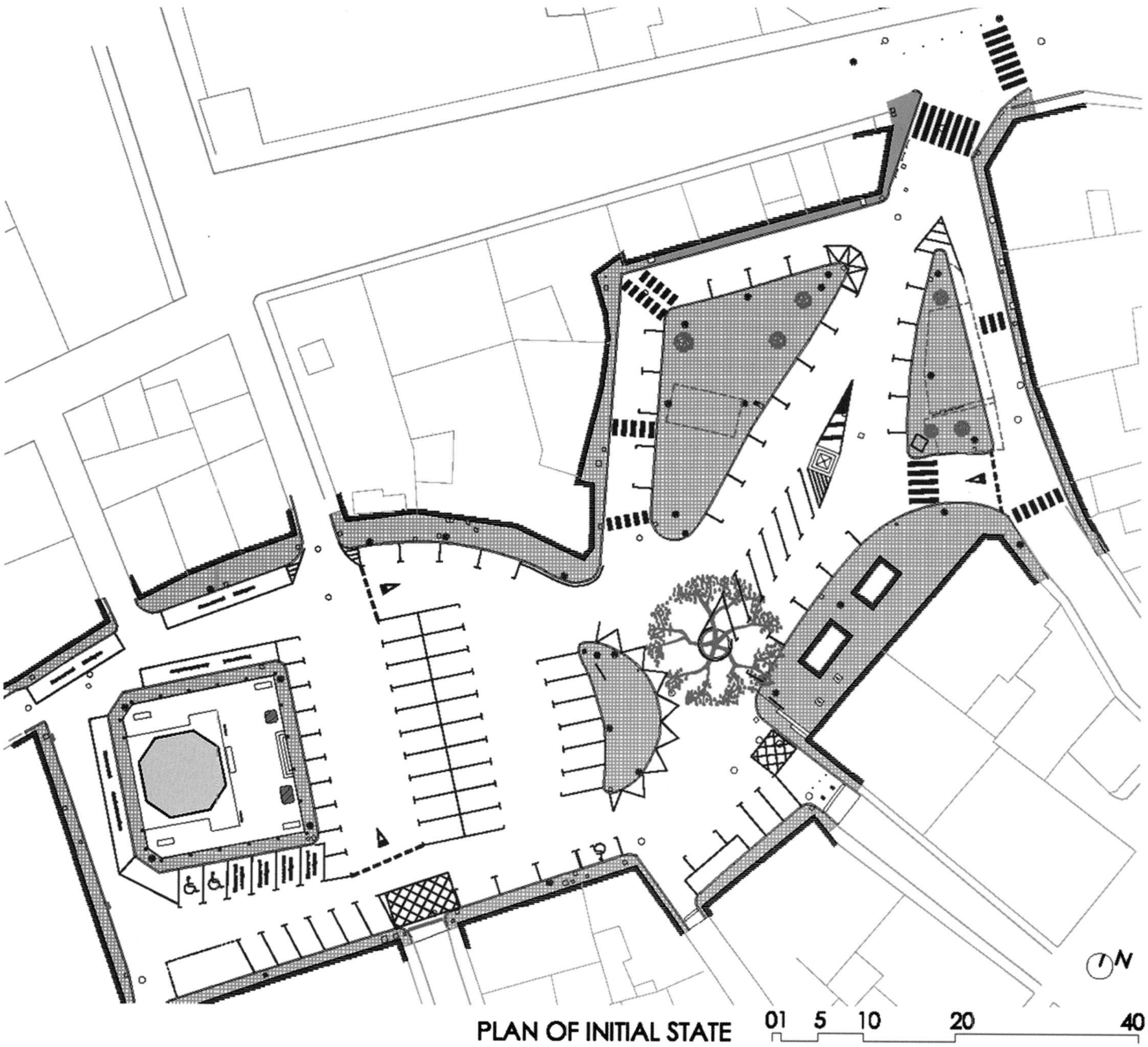

PLAN OF INITIAL STATE

The first step was to remove vehicular traffic by concentrating all the road connections and services on the southeast side of the squares, thereby freeing up a generously expansive central space for citizens' use.

The work has to be conducted at ground level due to the existence of an old cemetery, which prompted the design of two large beds for plants running lengthwise – each of them dubbed a "natural meander", and could be filled with sufficient earth for planting trees in.

Between them, they create a long urban feature that both protects the area from the traffic-bearing road and acts as a filter. Each stretch of the meander folds back on itself, embracing distinctive spots. Other areas include the various underground bins and banks for recycled waste and the loading and unloading parking places on the outer edge of the square.

The continuity of the meander is interrupted by transverse side entrances that facilitate access to the squares and connect them with their immediate surroundings.

The space is geometrically structured and arranged through the use of a single material for the paving (granite) throughout the square, the various areas within them being distinguished by distinctive formats:

– A perimeter strip paved with small slabs of granite that acts as the transition between the buildings, their entrances and the main central area of the squares.

– The central area, in which the slabs of granite are geometrically arranged and come in two different shapes.

The squares are now seen as a properly appointed place that encourages people to enjoy a quiet moment and to meet up with others; a peaceful space that allows the architecture of their surroundings to speak, the choir of traditional buildings that serve as their backdrop.

OVERALL PLAN OF THE WORK

1. Children's' playground area | 2. Natural meander | 3. Open central space for citizens' use for recreational, cultural and festive purposes | 4. Space for temporary stage | 5. Santa Maria Bell Tower | 6. Replanting of original trees | 7. Standardisation of cafés' awnings and sunshades | 8. Churro stalls | 9. Benches | 10. Perimeter strip paved with small granite slabs | 11. Herois Monument, a feature of the original square | 12. Old tree

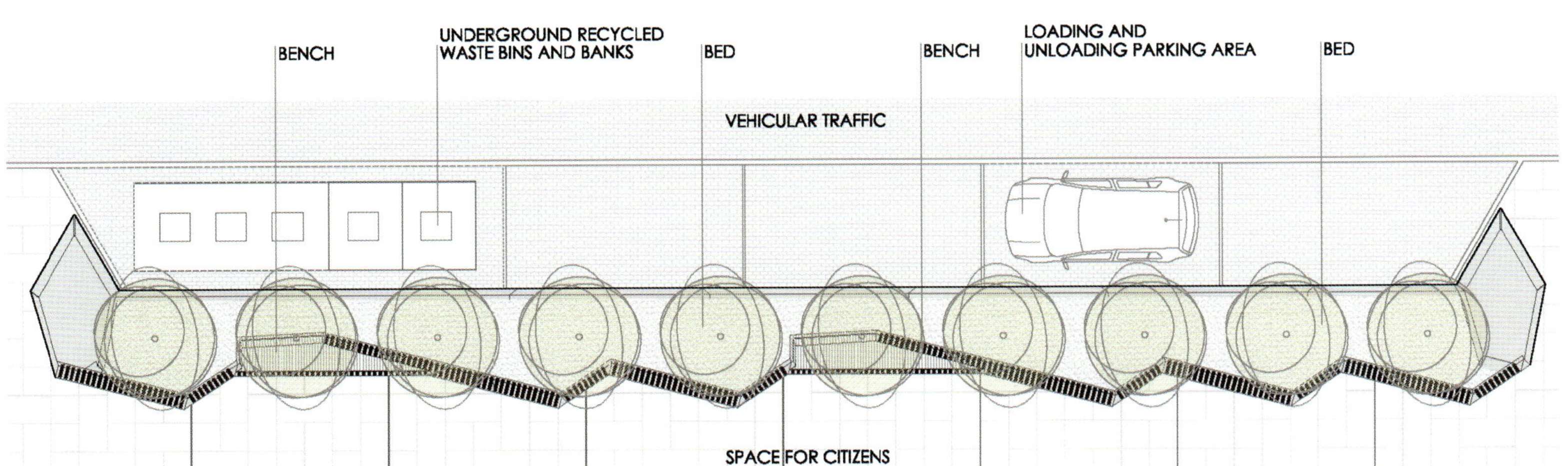

PLAN OF NATURAL MEANDER

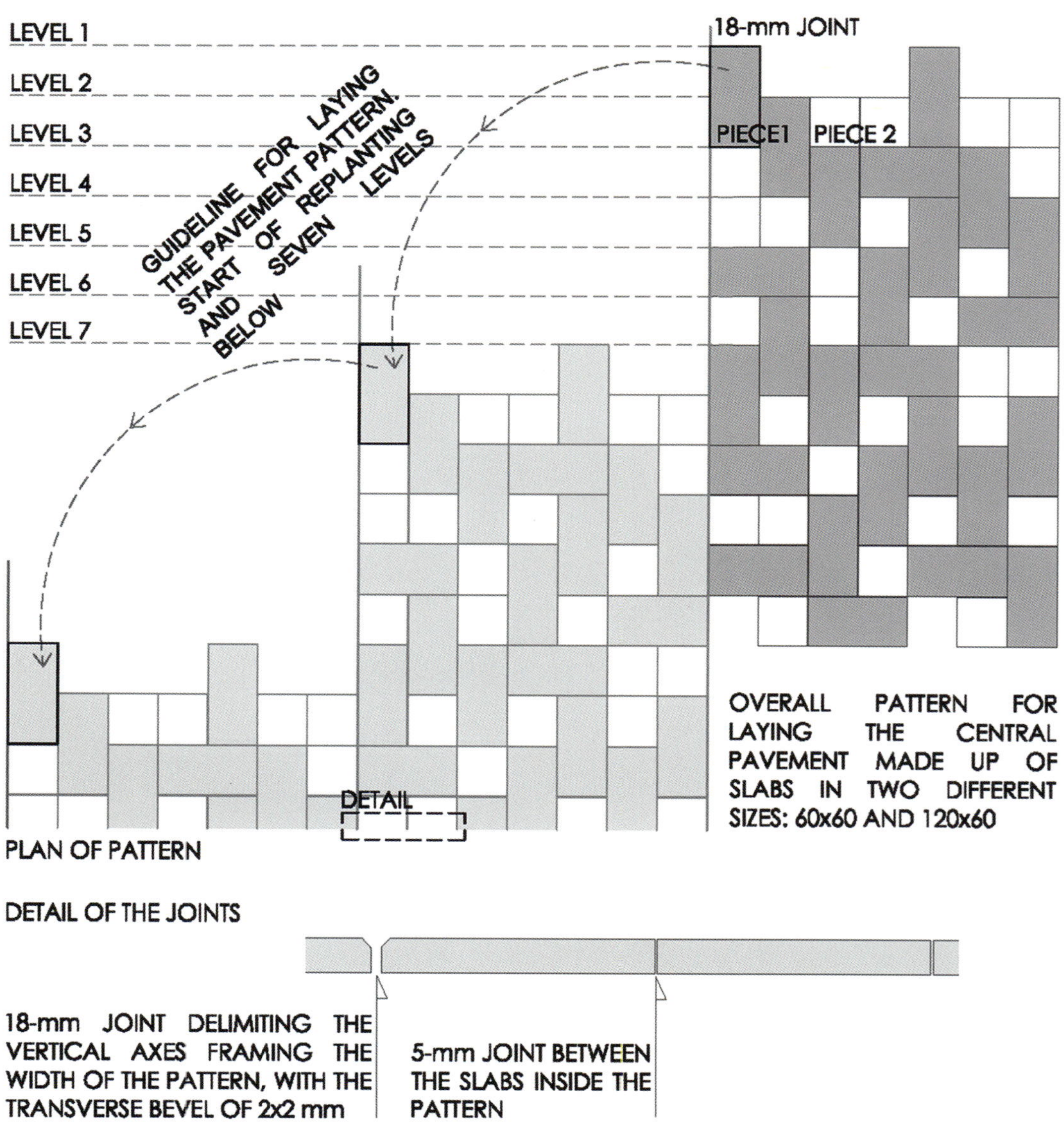

DETAIL OF PAVING

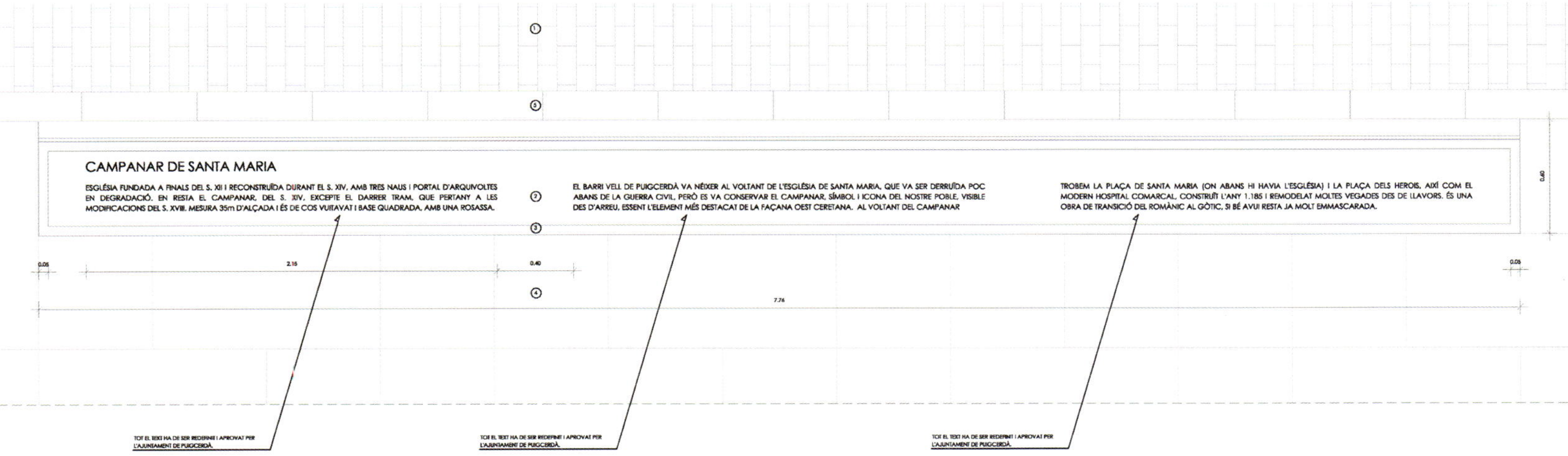

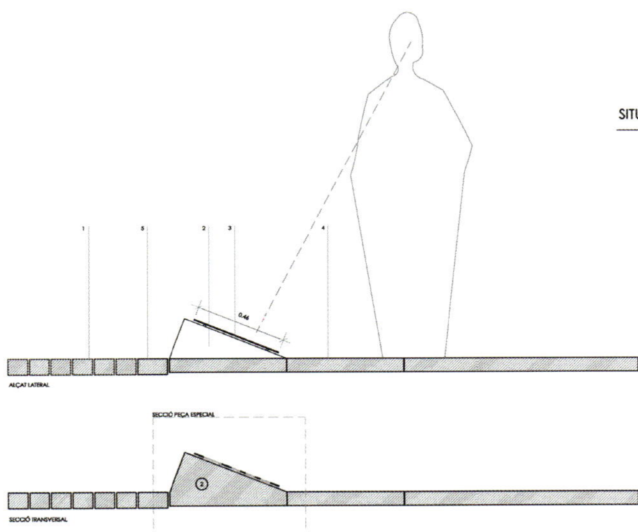

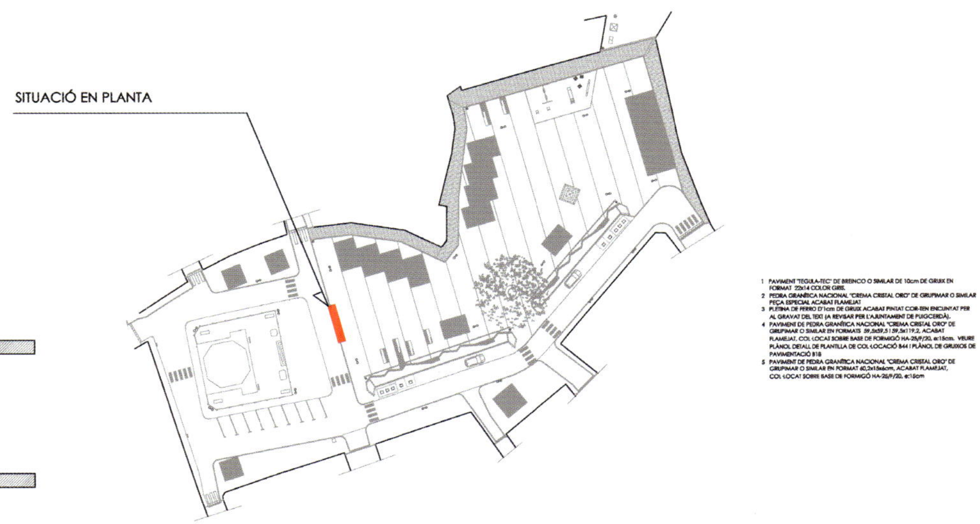

普格瑟达广场工程的第一阶段是拆除原有的车辆通道，将所有的道路交通和服务设施都设置在广场的东南方，这样可以打造一个占地面积更广的公共区域供市民使用。

广场项目首先从地面工程开始，需要对原有的古老墓地进行处理，按照规划这里将重新填满肥沃的土壤，形成两条纵向分布的自然绵延的绿化带。

两条绿化带中间形成了一条长长的具有城市特点的区域，遮蔽了交通车辆带来的噪音，并在一定程度上起到了过滤的作用。两条绿化带各自延伸并折回，形成两个与众不同的闭合区域。其他空间还包括多个用来堆积废弃物的地下储藏室和进出广场外侧停车场的通道。

绿化带的一段是一条通向侧面入口的小路，方便公众进出广场，并直接通向周边的其他区域。

广场的整个空间都使用花岗岩石板铺设，以不同的几何造型对各区域进行区分：设计师们在中心区域周边铺设小花岗岩石板，作为建筑、入口以及中心区域的边界；广场中心区域铺设的花岗岩石板巧妙地构成了两个不同的形状。

现在的广场已经成为公众按时聚会，并享受静谧时光的地方；精心的设计和规划让广场更加安静平和，仿佛时刻在与周围默默交流，古老的建筑也仿佛在轻声吟唱，共同维护着这个清新美好的公共环境。

NOVA

CASINO CERETA

TJUVHOLMEN

Oslo, Norway

图霍尔曼公园

Tjuvholmen was originally one of the islands in the Oslo Fjord but was annexed to the mainland and incorporated into the dock and harbor area about 100 years ago. Tjuvholmen, with close contact with the sea, will be re-established in a long coastline created by the new canals and public spaces and a lush park to be located along the shoreline. In addition to these opportunities for city life and recreation there will be shops, restaurants, an activity center with art museum and a sculpture park to attract the public to Tjuvholmen, both in summer and winter.

Tjuvholmen consists of three areas: Odden, Holmen and Skjaeret, and Bjørbekk & Lindheim are responsible for the concept and design of the landscaping of outdoor areas at Odden.

Above ground the district is car-free although there is parking for approx. 1,000 cars in the underground parking garages. The street network of the district is laid out in a fan shape creating changing patterns of sunshine throughout the day. The terrain on Odden has deliberately been created with the highest point at the central square. Three water features and a central tree create character with water features representing water's different characteristics. Water runs from the fountains in channels through the streets towards the Fjord.

• **Team** - Jostein Bjørbekk, Rune Vik, Håvard Strøm, Sigrid Lofthus Drange, Aina Skjærvø, Christer Ohlsson

• **Area** - 26,000 m^2 • **Photographer** - Bjørbekk & Lindheim AS

UNION
UNION GRUPPEN
BØLGEN & MOI
BØLGEN & MOI

图霍尔曼地区起初是奥斯陆海峡中的一个小岛，一百年前为大陆所兼并，划归为码头和海湾区域。这里毗邻大海，在长长的海岸线旁边新建了运河和其他公共区域。项目计划对紧邻海岸线的海滩地段进行重建，沿海岸线营建一个绿树如茵的公园。另外，项目中还包括商铺、餐馆、包括艺术博物馆和雕塑公园在内的娱乐中心，完工后还将为市民们提供一个综合休闲娱乐的场所，一年四季吸引市民和游客光顾图霍尔曼。

图霍尔曼地区由奥登、霍尔曼与斯凯瑞特三部分构成，博约贝克与林德海姆设计工作室负责奥登地区项目的室外景观设计方案。

奥登地区设有一个可容纳约1000辆汽车的地下停车场，地上则禁止车辆通行。奥登地区的道路规划整体呈扇形，白天阳光充足。这里地势最高，是整个公园的中心，植有一棵大树，旁边是三个不同的人造水系景观，形成了一道独特的风景。喷泉的水沿着水系通道，穿过街区，直达海边。

Brighton, England

NEW ROAD, BRIGHTON AND HOVE

布莱顿霍夫的新建大街

New Road in Brighton is now an important street in the city center's Cultural Quarter. It is situated outside the internationally recognized architectural icon of the Brighton Pavilion and is lined with the main theatres and performance venues.

Landscape Projects, working with Gehl architects from Copenhagen, were commissioned by Brighton and Hove City Council to transform New Road into a pedestrian friendly place which would be a focus of the Brighton's Cultural Quarter.

The design of the street includes a continuous level surface made of granite, chosen from a variety of tones to create a lively, subtly patterned street. Granite edgings and channels in contrasting white granite define site features. Street signage and clutter is reduced to a minimum, and includes massive stone bollards and timber benches.

Since completion, footfall has increased over 600% and staying activity by 200%; it is one of the four most popular places for people to spend time in the city. The New Road project has transformed this part of Brighton, making a socially and environmentally sustainable public realm place.

- **Collaborator** - Gehl architects (Architect), Stockley (Engineers), Simon, Fenton Partnership (QS), Esther Rollinson (Artist)
- **Project Management** - Jim Mayor at Brighton and Hove City Council
- **Area** - 2,800 m^2
- **Photographer** - Shaw & Shaw, Landscape Projects

ROYAL
Theatre Royal Brighton
THEATRE ROYAL

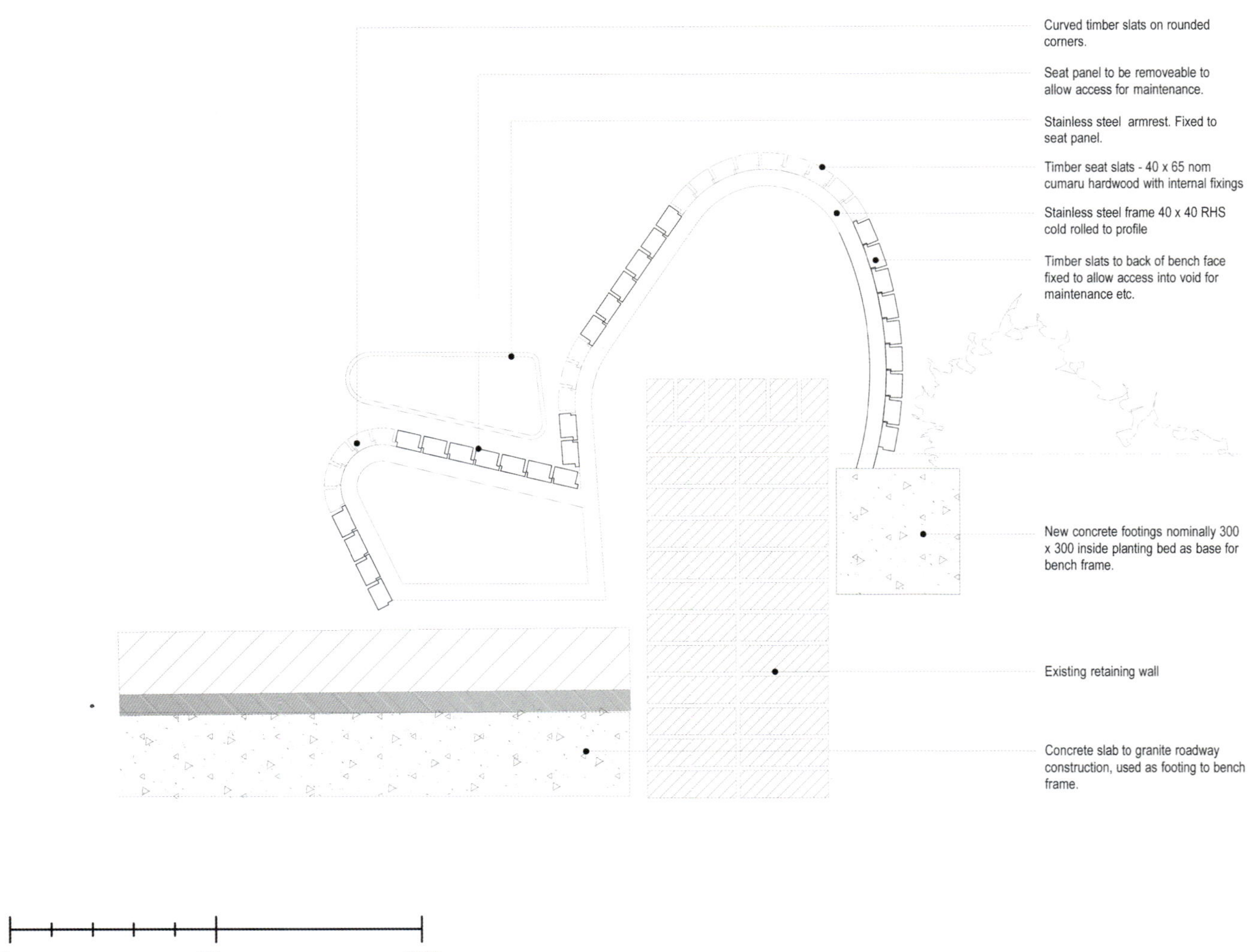

Curved timber slats on rounded corners.
Seat panel to be removeable to allow access for maintenance.
Stainless steel armrest. Fixed to seat panel.
Timber seat slats - 40 x 65 nom cumaru hardwood with internal fixings
Stainless steel frame 40 x 40 RHS cold rolled to profile
Timber slats to back of bench face fixed to allow access into void for maintenance etc.
New concrete footings nominally 300 x 300 inside planting bed as base for bench frame.
Existing retaining wall
Concrete slab to granite roadway construction, used as footing to bench frame.
0mm
500mm
1000mm

JENKINS
LET BY
Theatre Royal Brighton
ROYAL

Section DD

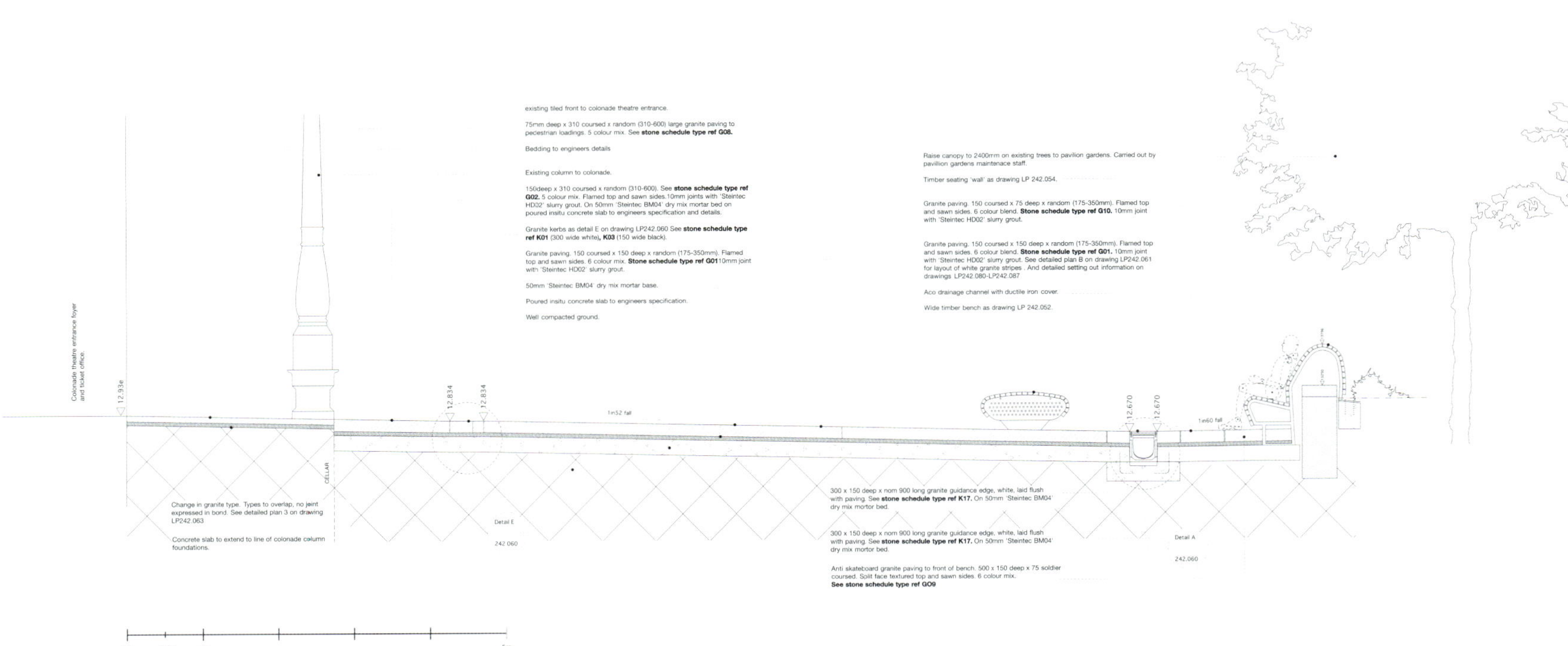
Colonade theatre entrance foyer and ticket office.
12.93e
existing tiled front to colonade theatre entrance.
75mm deep x 310 coursed x random (310-600) large granite paving to pedestrian loadings. 5 colour mix. See **stone schedule type ref G08.**
Bedding to engineers details
Existing column to colonade.
150deep x 310 coursed x random (310-600). See **stone schedule type ref G02.** 5 colour mix. Flamed top and sawn sides.10mm joints with 'Steintec HD02' slurry grout. On 50mm 'Steintec BM04' dry mix mortar bed on poured insitu concrete slab to engineers specification and details.
Granite kerbs as detail E on drawing LP242.060 See **stone schedule type ref K01** (300 wide white)**, K03** (150 wide black).
Granite paving. 150 coursed x 150 deep x random (175-350mm). Flamed top and sawn sides. 6 colour mix. **Stone schedule type ref G01**10mm joint with 'Steintec HD02' slurry grout.
50mm 'Steintec BM04' dry mix mortar base.
Poured insitu concrete slab to engineers specification.
Well compacted ground.
Raise canopy to 2400mm on existing trees to pavilion gardens. Carried out by pavillion gardens maintenace staff.
Timber seating 'wall' as drawing LP 242.054.
Granite paving. 150 coursed x 75 deep x random (175-350mm). Flamed top and sawn sides. 6 colour blend. **Stone schedule type ref G10.** 10mm joint with 'Steintec HD02' slurry grout.
Granite paving. 150 coursed x 150 deep x random (175-350mm). Flamed top and sawn sides. 6 colour blend. **Stone schedule type ref G01.** 10mm joint with 'Steintec HD02' slurry grout. See detailed plan B on drawing LP242.061 for layout of white granite stripes . And detailed setting out information on drawings LP242.080-LP242.087
Aco drainage channel with ductile iron cover.
Wide timber bench as drawing LP 242.052.
12.834
12.834
1in52 fall
12.670
12.670
CELLAR
Change in granite type. Types to overlap, no joint expressed in bond. See detailed plan 3 on drawing LP242.063
Concrete slab to extend to line of colonade column foundations.
300 x 150 deep x nom 900 long granite guidance edge, white, laid flush with paving. See **stone schedule type ref K17.** On 50mm 'Steintec BM04' dry mix mortor bed.
300 x 150 deep x nom 900 long granite guidance edge, white, laid flush with paving. See **stone schedule type ref K17.** On 50mm 'Steintec BM04' dry mix mortor bed.
Anti skateboard granite paving to front of bench. 500 x 150 deep x 75 soldier coursed. Split face textured top and sawn sides. 6 colour mix.
See stone schedule type ref GO9
0m
0.5m
1m
5m

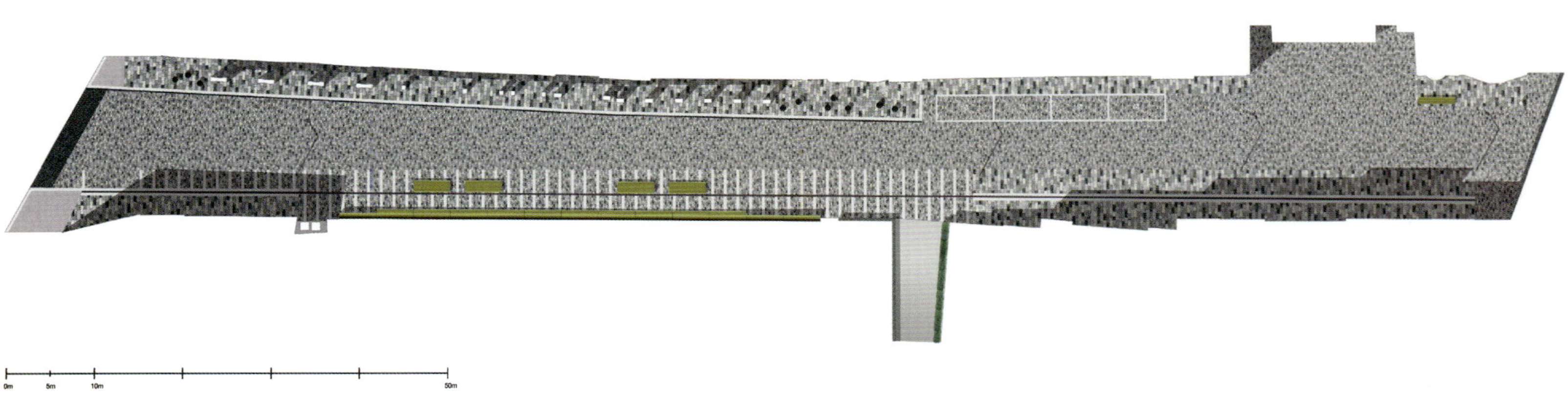
0m
5m
10m
50m

布莱顿霍夫的新建大街在市中心文化区占有重要的地位。新建大街位于举世闻名的地标性建筑布莱顿会展中心外侧，临街还有市区主要的剧院和艺术表演场馆。

布莱顿霍夫市政委员会经过协商，委任园林景观项目组联合来自哥本哈根的吉尔建筑工作室，将新建大街设计改造为一个热情友好的步行区，使其成为布莱顿霍夫文化区的焦点。

整条大街的设计规划包括从大量材料中选择合适色调的花岗岩石板铺就地面，打造一条连续延伸的、充满活力的大街。花岗岩石板路的边缘和隧道使用白色花岗岩作为装饰，极富当地特色。利用巨大的石柱和实木长椅，将街区标志的数量降到最少。

工程完成以后，客流量增加了600%，行人驻足参与活动增加了200%；成为城市中最能吸引市民欢度时光的四个场所之一。新建大街工程改变了这一地区的社会形象，将这里打造为一个集社交和环境为一体的可持续性发展的公共区域。

• **Landscape Architect** - Matthias Funk, Hiltrud M. Lintel, Rainer Sachse • **Firm** - scape Landschaftsarchitekten GmbH

"NORTHERN CITY CENTER AND PEDESTRIAN AREA" KAMEN

Kamen, Germany

卡门市北中心广场及步行街区

The historic city center of Kamen – a city with 48,000 inhabitants, situated in the eastern part of the "Ruhrgebiet" – is a shopping and service location of more than local significance. The typical inhomogeneous building structure in Kamen with its historical and modern sequences of spaces and large pedestrian areas has a previously unused design potential. The aim of the redesign, carried out from the year 2006 to 2008 was to give Kamen's heterogeneous center a new identity and unity. A strikingly new and timeless quality in form and material sets urban design accents and contributes to the positive image of the city.

The design concept emphasizes the existing linking network of spaces. Linear spaces alternate with squares. The three spatial sequences West Street – Plane Square – West Street, Willy-Brandt-Square – Market Street – Market Square at the Evangelical Church and the spatial sequence in Adenauer Street are respectively defined according to their different characters. In contrast to the hard landscape of the shopping street, Adenauer Street is transformed to a "green boulevard" with a promenade and recreation area under trees. As a link between these two axes serves the north-south connection Market Street/ Willy-Brandt-Square. Willy-Brandt-Square is the focal point within this spatial sequence.

A paving strip forming a continuous characteristic carpet through the entire pedestrian zone unites the historical and modern architectural structure. Bandings in the pavement emphasize like "marquetries" special areas as for example recreation areas. A differentiated use of trees creates contrasting spaces.

Green squares invite children to play and form attractive entry spaces. At the most important landmark of the historic West Street – the Plane Square, a simple raised fountain is placed as a game with the existing topography. It provides a space for recreation close to the water and water games and is used as a terrace for outdoor dining.

• **Area** - 22, 000 m² • **Photographer** - Hiltrud M. Lintel

munkes
Hörgeräte
munkes Hörgeräte

Eschen-Allee
bestehende Platane
Edelkirchenhof
Wasserspiel
Fahrrad-
Baumschaukel
Spiel- und Sitztreppe
Sitzbank
Rutsche
Sandspiel
Rasenplateau
Parkplätze
Zufahrt Tiefgarage
Kämerstraße

Naturstein-
Einfassung
Steinbänke
Pflaster-Intarsie
Weststraße
Kleinpflaster
im Traufbereich

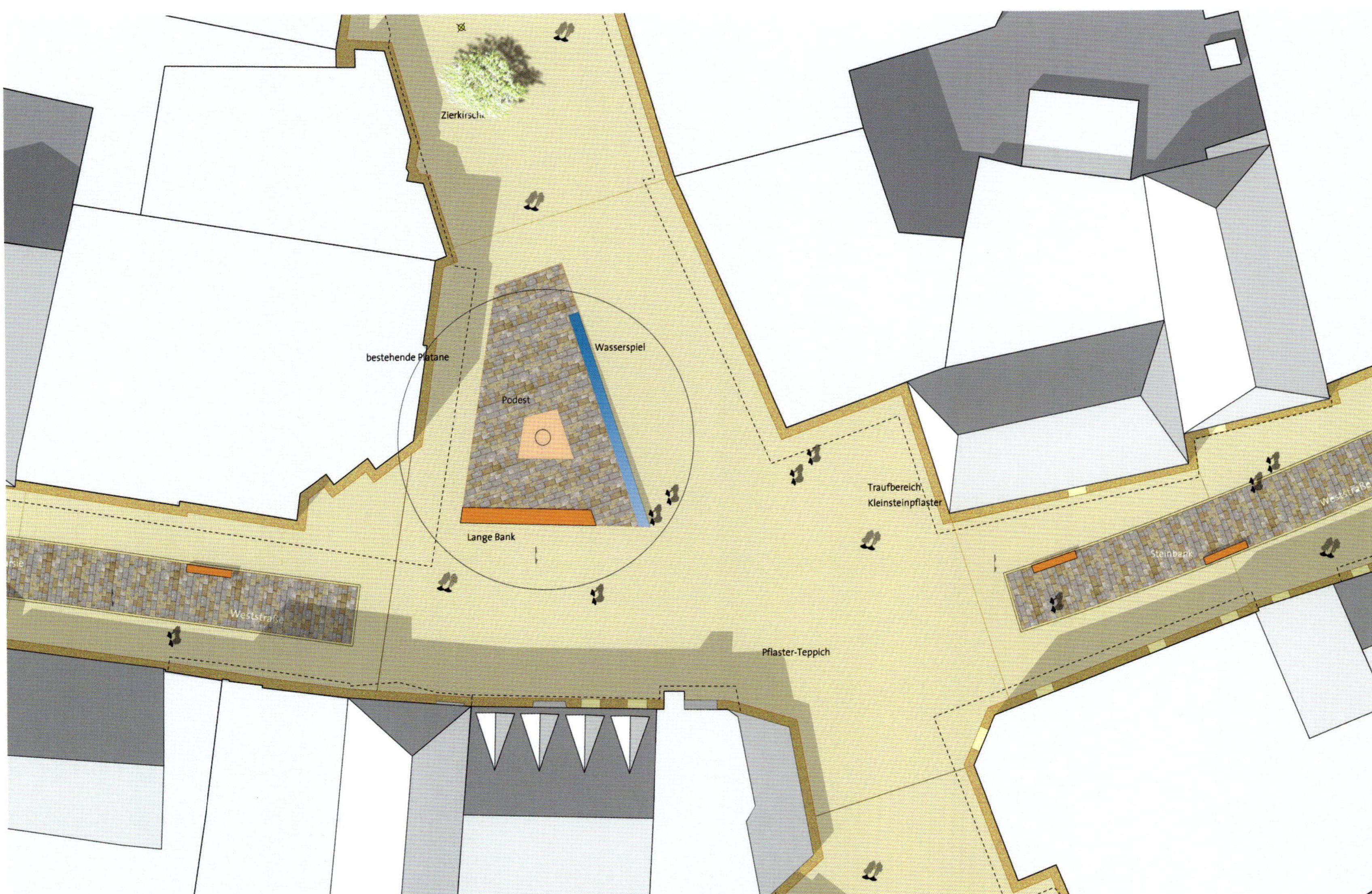
Zierkirsch
bestehende Platane
Wasserspiel
Podest
Lange Bank
Traufbereich,
Kleinsteinpflaster
Steinbänke
Weststraße
Pflaster-Teppich

卡门位于鲁尔区的东部，拥有 48000 名居民，其市中心区域是商业和服务中心，已远远超出了其本身的意义。卡门历史悠久，建筑结构多样，在城市建设的进程中留下了大量的建筑空地和步行区域，设计开发潜力巨大。2006 年至 2008 年进行的重新规划的目的是对鱼龙混杂的卡门市中心区域进行整合，打造一个崭新的城市风貌。设计规划的形式和选材在保证质量的基础上，更要给人一种耳目一新的感觉，奠定城市规划的基调，有助于树立城市建设的良好形象。

项目设计的理念主要是强调现有空间之间的相互连通。广场内形成三个线性空间交替的布局，分别是西街—平面广场—西街，福音教堂附近的威利布兰德广场—市场街—市场广场和阿登纳街区，三个线性空间遥相呼应，特点鲜明。与商业街区的硬质景观相比，阿登纳街区被规划为一条绿色的林荫大道，两侧是供市民娱乐漫步的休闲区域。在三个线性空间之间，有一条横贯市场街和威利布兰德广场的南北向通道，这样威利布兰德广场就成为了整个项目的焦点。

设计师规划了一条长长的石板路，如同一条独特的地毯横穿整个步行街区，将具有历史风貌的街区与现代化的建筑结构连接在一起。包括休闲区域在内的特殊功能地块，如同镶嵌在步行区域当中，十分醒目。广场中利用各种不同树种，根据不同的用途，隔开很多小的空间。

绿树如茵的广场吸引孩子们来这里嬉戏玩耍，同时，形成了一个极具吸引力的入口空间。极具历史意义的西街—平面广场区域是最重要的地标性地段，设计师们在这里设计了一个简单的喷泉，搭配周边现有的地势和建筑风格，为市民在休闲之余提供了一个亲临水系的游戏场所，一个可以享用室外餐饮的平台建筑。

APOTHEKE
REWE

• **Architect** - Vibeke Rønnow • **Firm** - C. F. Møller Architects • **Area** - 1,500 m²

DANISH CROWN-AREAS, NR. SUNDBY BRYGGE

Aalborg, Denmark

丹麦科王集团分区，布鲁日 NR. 松德比

The 120,000 square meters former Danish Crown abattoirs in Nr. Sundby are being transformed into a modern, mixed-use urban area, tied together by a pedestrian zone, squares and green parks. C. F. Møller Architects is responsible for the masterplan and design guidelines, focusing on the connecting central urban "rambla" as the main axis linking the city and harbour front.

The pedestrian street is designed around a small stream, springing from the surplus of groundwater, and running through the street in an open rill to the firth. Along the way, the stream runs through mossy pools, and a larger cor-ten steel mirror-pool. Surfaces and pavings are of honey-coloured concrete and slate, and are lit by integrated LED-lighting sources.

位于松德比的12万平方米的前丹麦科王集团屠宰场将变成一个现代化的多功能市区，与之并列的还有行人专用区、广场和青翠的公园。C. F. Møller 建筑事务所负责总体规划和设计指导方案，重点将放在作为连接城市和港湾地区的主轴枢纽地带的“兰布拉”上。

步行街设计在小溪周边，溪水从丰盈的地下水中源源不断地涌出，在一条开放式的水沟中流淌，穿过街道，一直注入到河口湾。沿途，小溪穿过满是苔藓的水池和一个更大的耐候钢围成的明澈如镜的水池。步行街和小路是用蜂蜜色的混凝土和板岩铺设而成，照明选择的是集成的LED光源灯。

• **Landscape Architect** - Hanneke Kjne, Petrouschka Thumann, Han Konings, Remco Rolvink, Ronald Bron, Hilke Floris • **Firm** - HOSPER

GENK C-M!NE

比利时亨克 C-M!NE 广场

Genk, Belgium

Space for the spectacular!

The C-m!ne square, situated on a former coal mining site, is the central open space of the new cultural center of Genk.

One level surface

The need to create opportunities for a wide range of activities on the square results in a "one level surface". An obstacle-free surface ensures that the square can be used for a wide variety of purposes.

Building bricks

The square is paved with black slate slabs of different sizes and laid in an informal pattern. The black slate refers to the "black gold" from the mines and is the same material as the waste material from the mining activities that was placed on the terrins (cairns) around Genk. The paving includes lighting in the surface as well as the possibility for a water surface, the creation of mist just above the surface and removable seating. A great deal of attention is given to the night-time appearance of the square, with lighting illuminating the surrounding facades and the former mining shaft towers. The two shaft towers are both given a function on the square. The Belgian office NU architectuuratelier designed an attractive route that follows the former mining corridors under the oldest shaft tower and the ruin of the former mining reception building and ends with a fantastic view on top of the youngest and tallest shaft tower.

Single scatter street furniture

The chairs and stools, which are made of folded stainless steel plates, glitter like diamonds against the black surface of the square. The internal and rear surfaces of the furniture are powder-coated in fire-engine red.

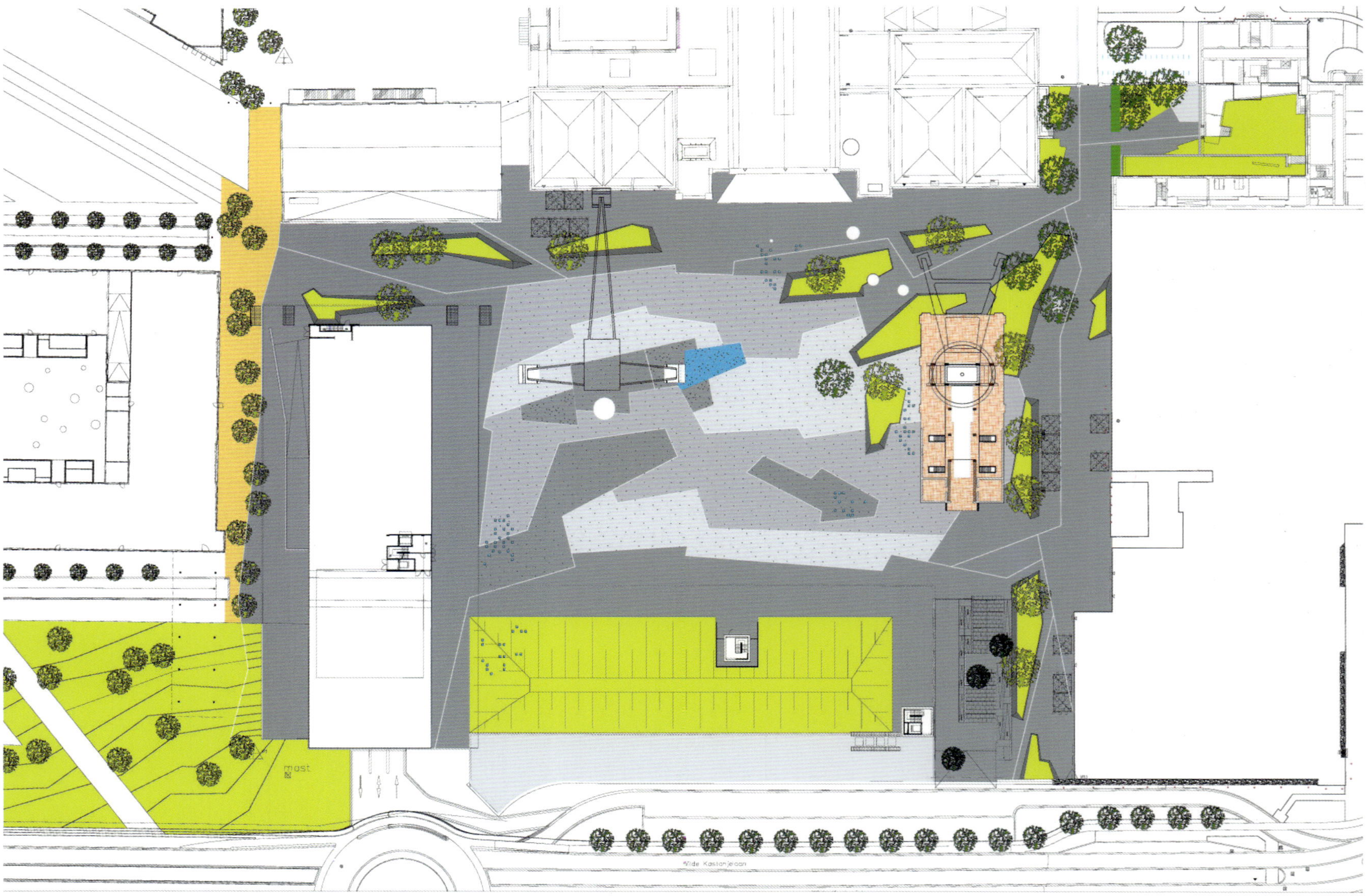

• **Partner** - ARA Atelier Ruimtelijk Advies,Carmela Bogman industrial design, Painting with Light, NU architectuuratelier

• **Area** - 5,000 m^2 • **Photographer** - Pieter Kers, HOSPER • **Client** - City of Genk

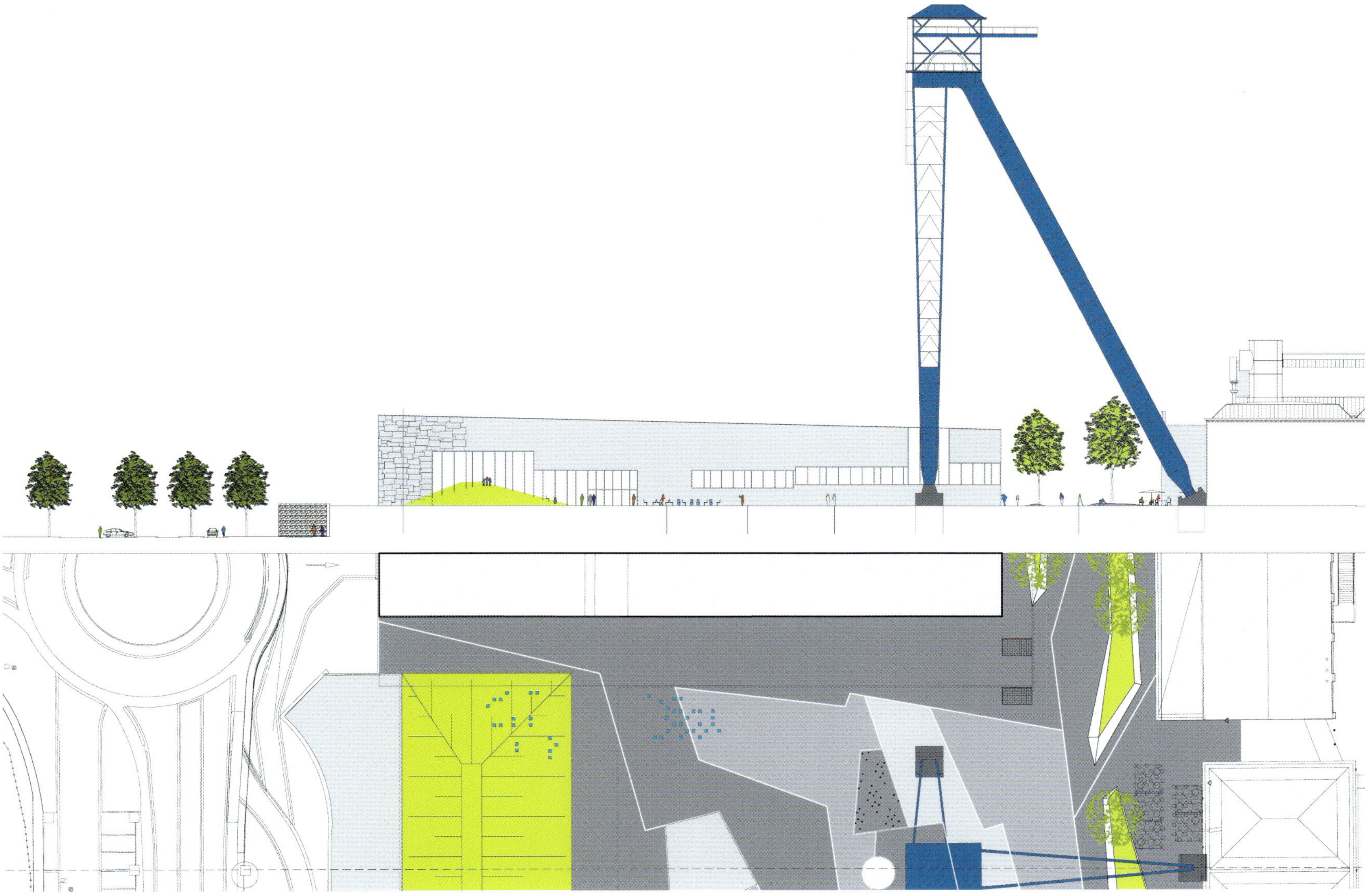

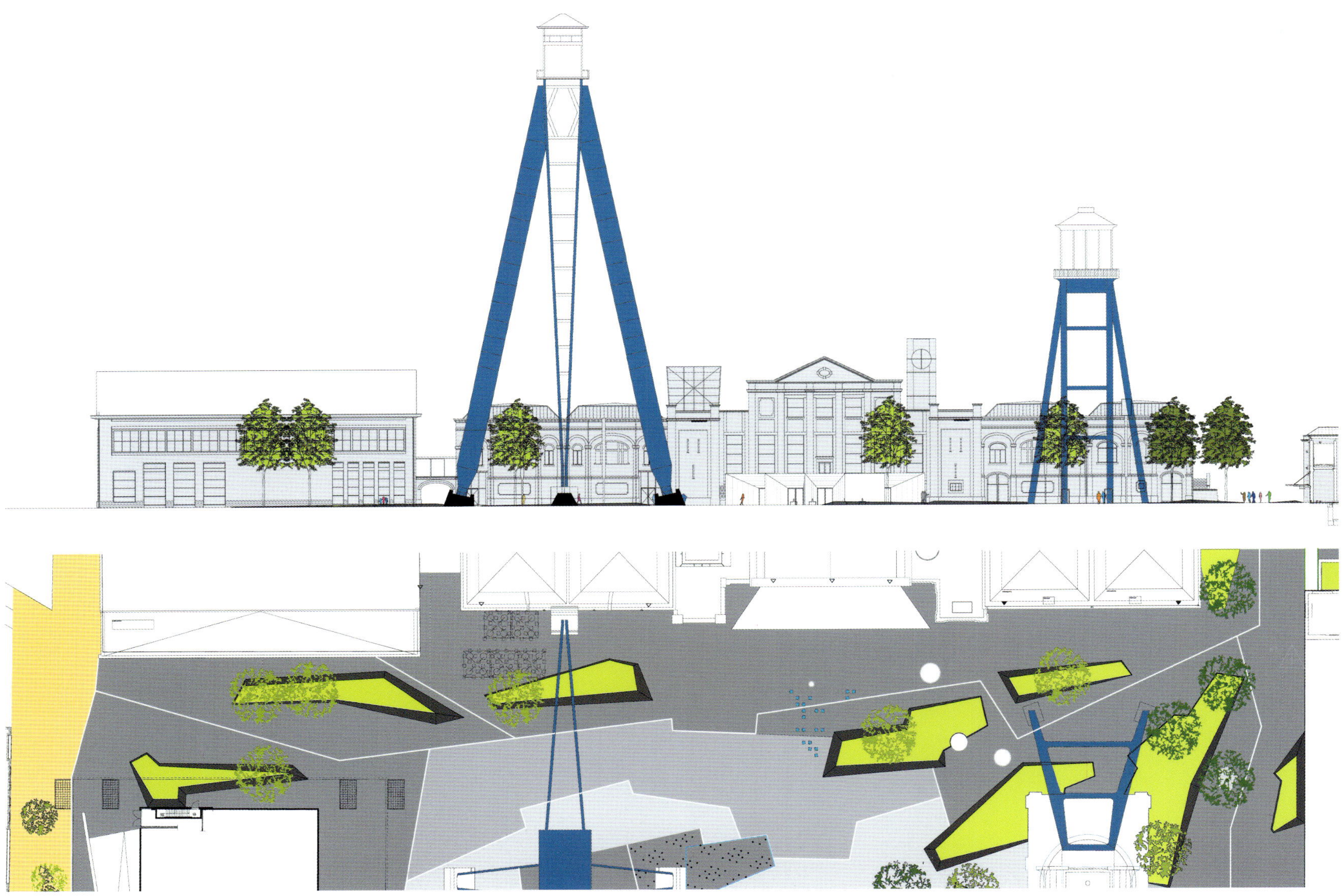

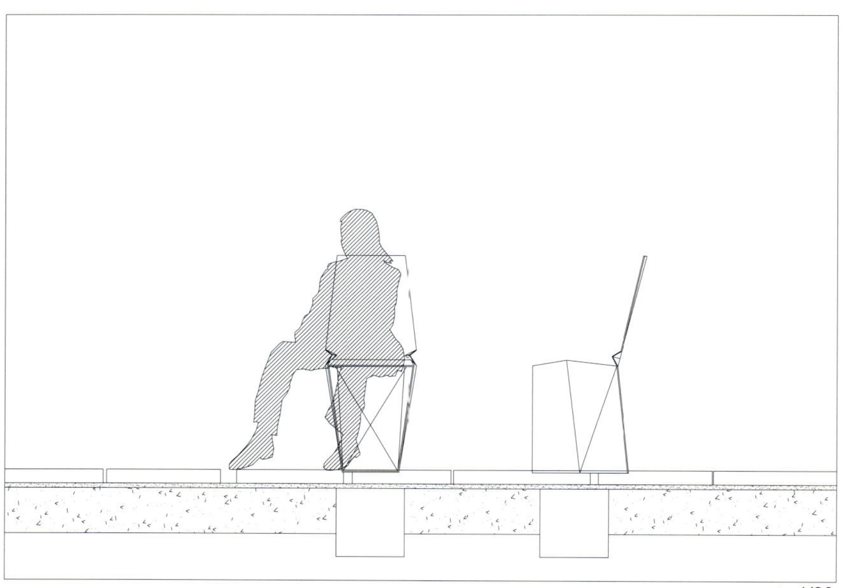

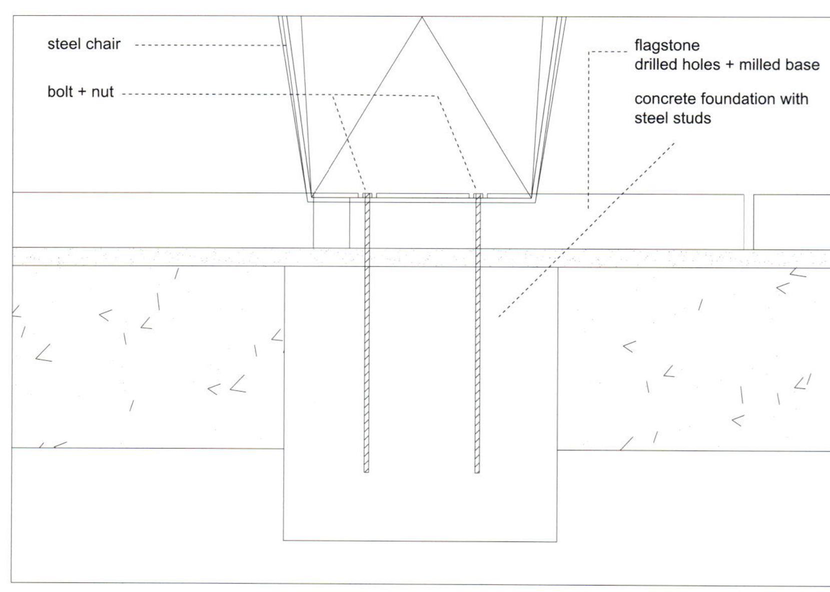
steel chair
bolt + nut
flagstone
drilled holes + milled base
concrete foundation with
steel studs

scale 1/20

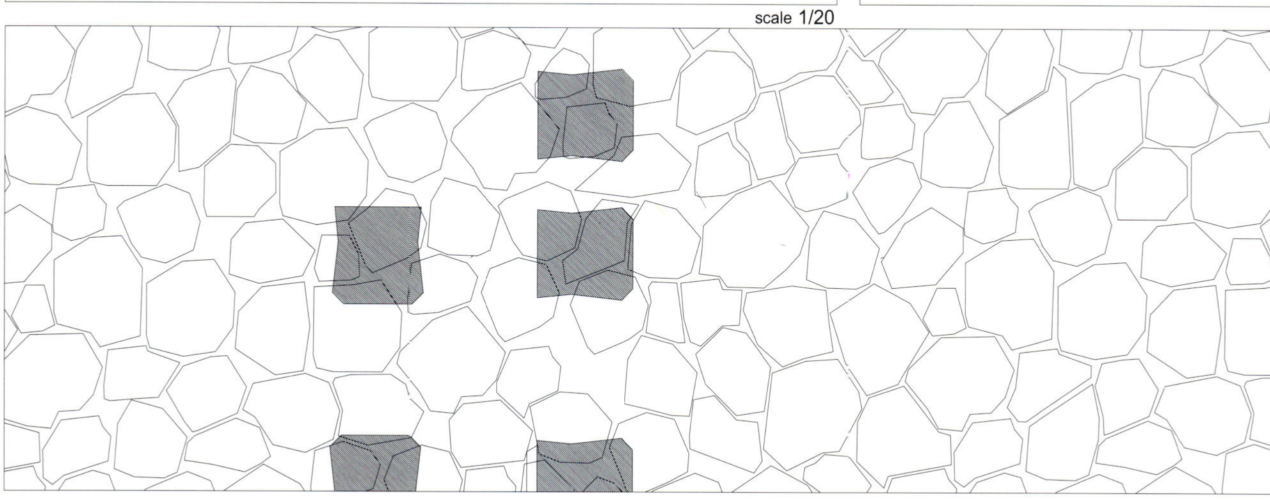

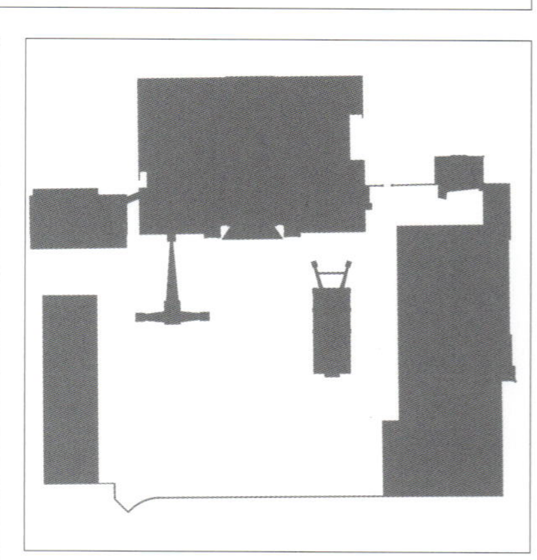

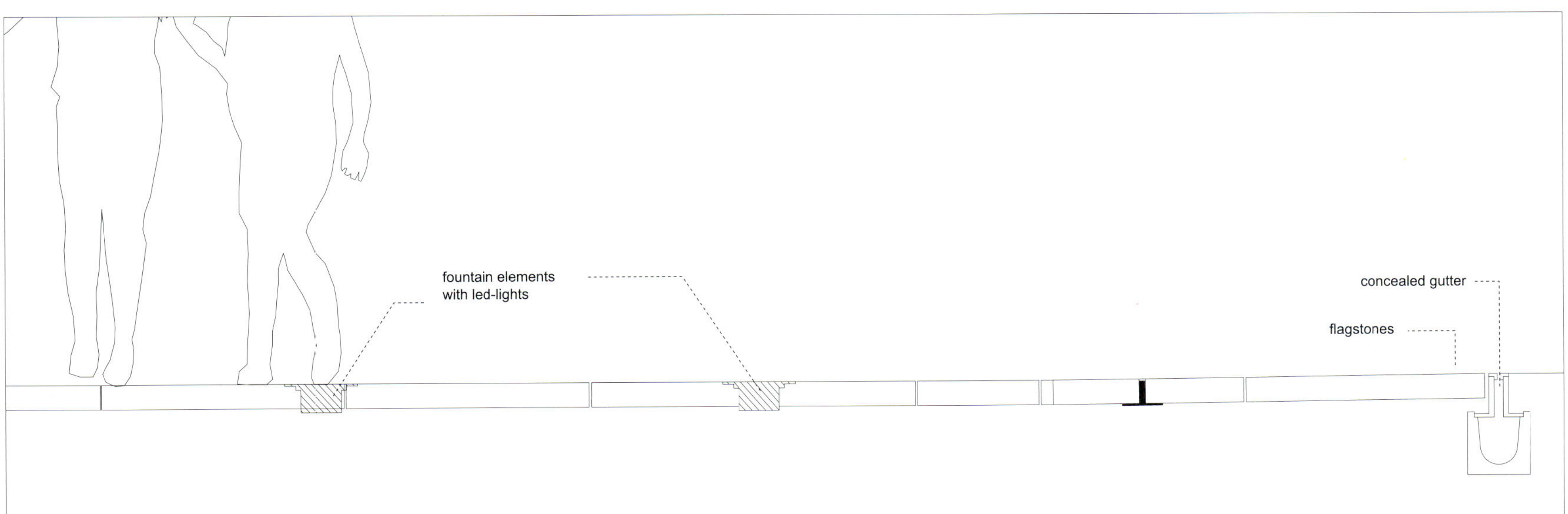
fountain elements
with led-lights
concealed gutter
flagstones

scale 1/10

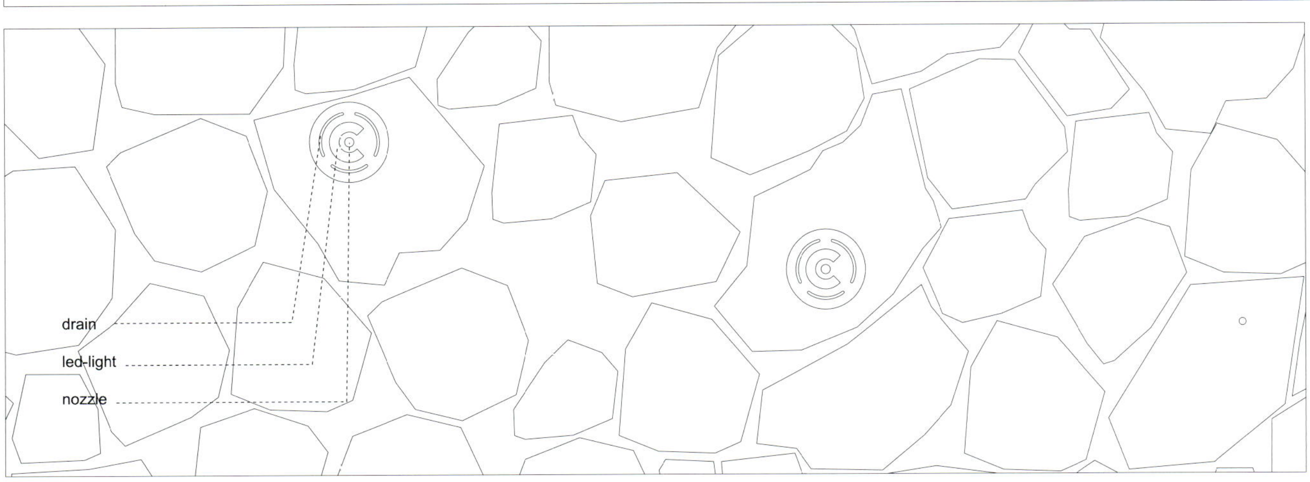
drain
led-light
nozzle

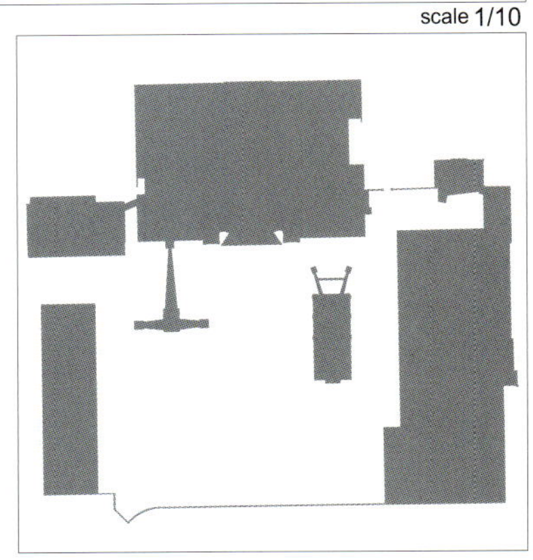

场面壮观的空间!

亨克市 C-m!ne 广场过去是一个采煤场，现在是亨克市新的文化中心所在地。

地表一层空间

为了便于举办大型的公众活动，最终将广场设计为只有地表一层空间。这样无障碍的广场表面设计，保证了广场广泛的应用范围。

砖砌建筑

整个广场由不规则形状的黑色石板铺成。黑色石板由之前矿物开采剩余的废料加工而成，象征着遍布亨克地区的矿物开采中的“黑金”。广场石板路表面平滑如水，可以反射光线，由反射光形成的薄雾经常悬浮在广场可移动座椅的上方。相对而言，设计师们更注重广场夜间的效果规划，尤其是广场正面以及保存下来的采煤支架周围，到了夜晚灯火通明。比利时 NU 建筑事务所赋予了两个采煤支架新的功能，在最古老的采煤支架和采煤接待大楼的下方，利用原有的采煤通道，设计建造了引人注目的人行步道、直通最后建造的、高度最高的采煤支架顶端，那里拥有绝佳的视野，景色壮丽，令人难忘。

独立而分散的街边景物

由折叠不锈钢板制成的椅子和高脚凳，闪烁着钻石般的光芒，与广场漆黑的表面形成强烈的对比。景物的内侧和背面涂有消防红色的粉末状物质。

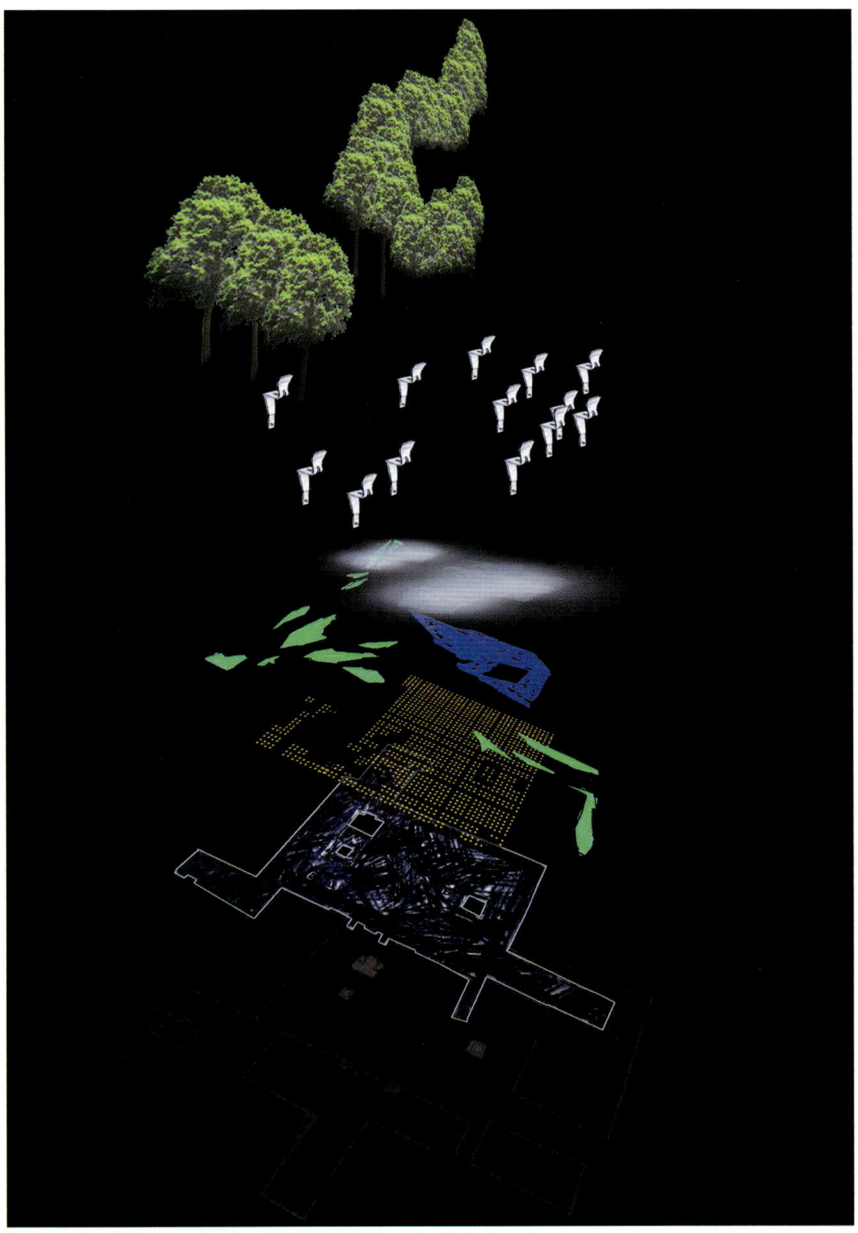

• **Landscape Architect** - Bureau B+B stedebouw en landschapsarchitectuur • **Area** - 40,350 m^2 • **Photographer** - Renee Klein, Frederica Rijkenberg

PUBLIC SPACE "BLOOMING" CITYCENTER NIEUWEGEIN

新维根市城市中心广场

Citycenter Nieuwegein, The Netherlands

The design proposes a dynamic whole, comprised of transparent storeys, in which continuous activity is possible from morning to night. The city center becomes outwardly open in a represen-ative manner and makes contact with the water of the Doorslag Canal. The public space is distributed over different levels, unified by means of a folded ground level that leads the visitor in a self-explanatory manner to all components of the programme. The elaboration of the design for the public space has as its concept "the blooming city". This translates into a striking pavement pattern employing natural stone in two different mixtures. The pattern breaks free of both plan and architecture by means of an abstract representation of such natural elements as branches and flowers. The greenery is concentrated on the plazas. Tree islands and raised tree-squares with integrated seating elements give each plaza its own unique character. Blossoming trees, such as magnolias and prunuses, alternating with trees that feature lovely autumn colours, give the new urban heart an attractive appearance throughout the year. One of the plazas owes its identity in part to its unusual fencing. Here, the floral pattern attains independence in the third dimension.

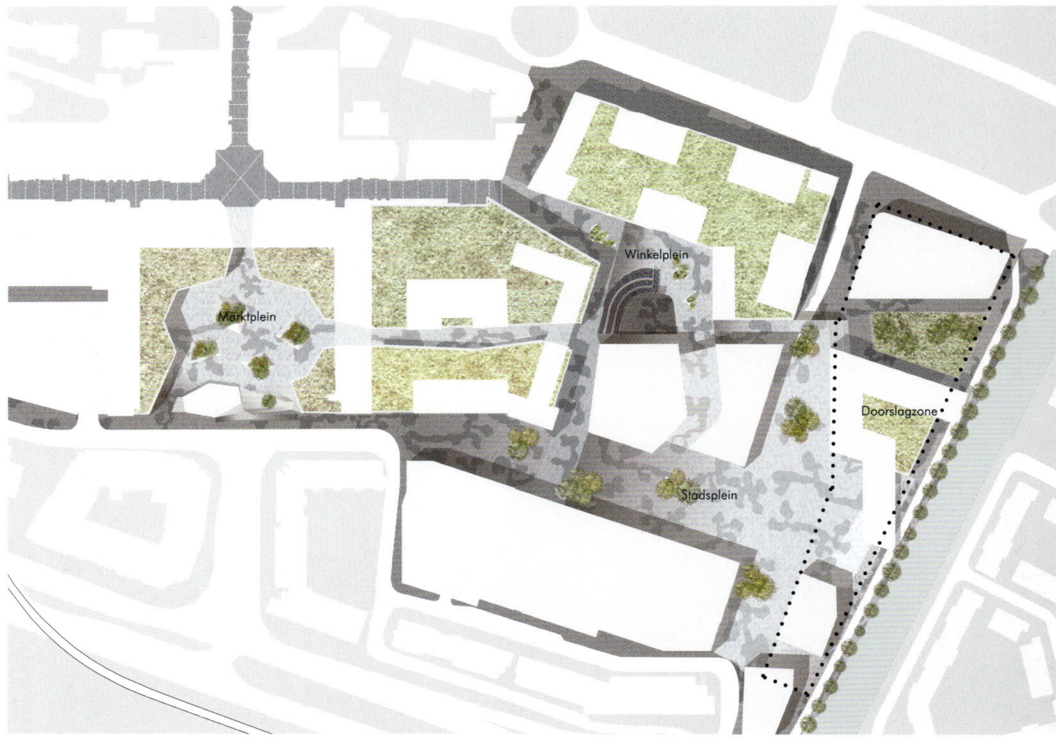

SATURN
SATURN POWER SERVICE
LCD•PLASMA•HIFI•DVD
FOTO•ENTERTAINMENT•TELECOM•WITGOED•MULTIMEDIA
SATURN
ACTIE!
ACTIE!
PIXIE
KOSTER

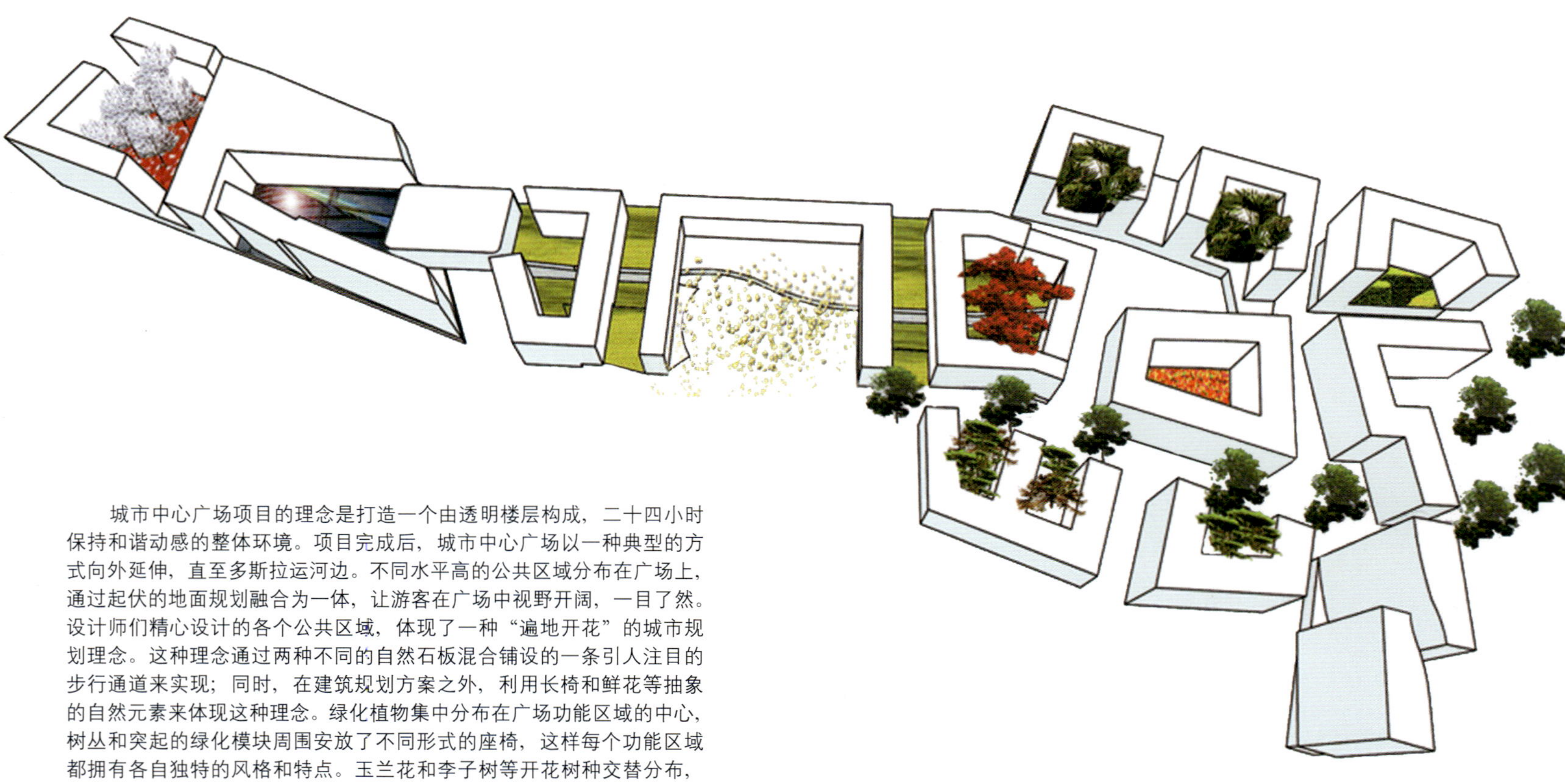

城市中心广场项目的理念是打造一个由透明楼层构成，二十四小时保持和谐动感的整体环境。项目完成后，城市中心广场以一种典型的方式向外延伸，直至多斯拉运河边。不同水平高的公共区域分布在广场上，通过起伏的地面规划融合为一体，让游客在广场中视野开阔，一目了然。设计师们精心设计的各个公共区域，体现了一种“遍地开花”的城市规划理念。这种理念通过两种不同的自然石板混合铺设的一条引人注目的步行通道来实现；同时，在建筑规划方案之外，利用长椅和鲜花等抽象的自然元素来体现这种理念。绿化植物集中分布在广场功能区域的中心，树丛和突起的绿化模块周围安放了不同形式的座椅，这样每个功能区域都拥有各自独特的风格和特点。玉兰花和李子树等开花树种交替分布，色彩斑斓，为这个新的城市中心增添了一道独具魅力的风景。广场中有一个单独的功能区域，周围设置了独特的围栏，围栏上的花卉图案呈现出完美的三维立体效果。

NATIONAL SEPTEMBER 11 MEMORIAL

New York, USA

美国 9.11 国家纪念碑广场

Through a collaborative process, PWP Landscape Architecture, Michael Arad (Handel Architects) and Davis, Brody, Bond built a memorial that commemorates the victims of the attacks of September 11, 2001, at the World Trade Center, Shanksville, Pennsylvania, the Pentagon, and the World Trade Center attack of February 26, 1993. Two fountain-lined voids, on the locations of the destroyed twin towers, and a surrounding forest of oak trees form the core of the rebuilt World Trade Center in New York City and provide a place for contemplation and remembrance within this revitalized urban center.

Despite its apparent simplicity, the Memorial is a massive green roof – a fully constructed ecology – that operates on top of multiple structures including the PATH station and tracks, the Memorial Museum, a central chiller plant, parking, and additional infrastructures. Visitors will leave the everyday life of the city and enter into a sacred zone defined by a dense forest of 416 oak trees. Above, the high canopy of leaves provides welcome shade in the heat of the summer and seasonal color in the fall. In the winter the sun will cast shadows through a light tracery of bare branches, and in spring, the trees will express the renewal of nature. In this way, the overwhelming losses of September 11th are given permanent presence. The 4,047 square meters voids of the fountains, cut over 9 meters into the site, are lined with waterfalls. Using full-scale mock-ups to study the performance of the water, the design team developed a tapered, rounded weir that is both water-and-energy efficient as well as highly visible and beautiful. With the addition of lighting, the waterfalls are also visible at night.

Throughout the design and construction process, sustainability of the plaza was considered in terms of both material endurance and landscape performance. The plaza surface-and-drainage infrastructure is designed to function as a large self-sustaining cistern. Water from rainfall and snow melt is channeled into large holding tanks and re-used to support the Memorial forest via a specialized drip-and-spray irrigation system.

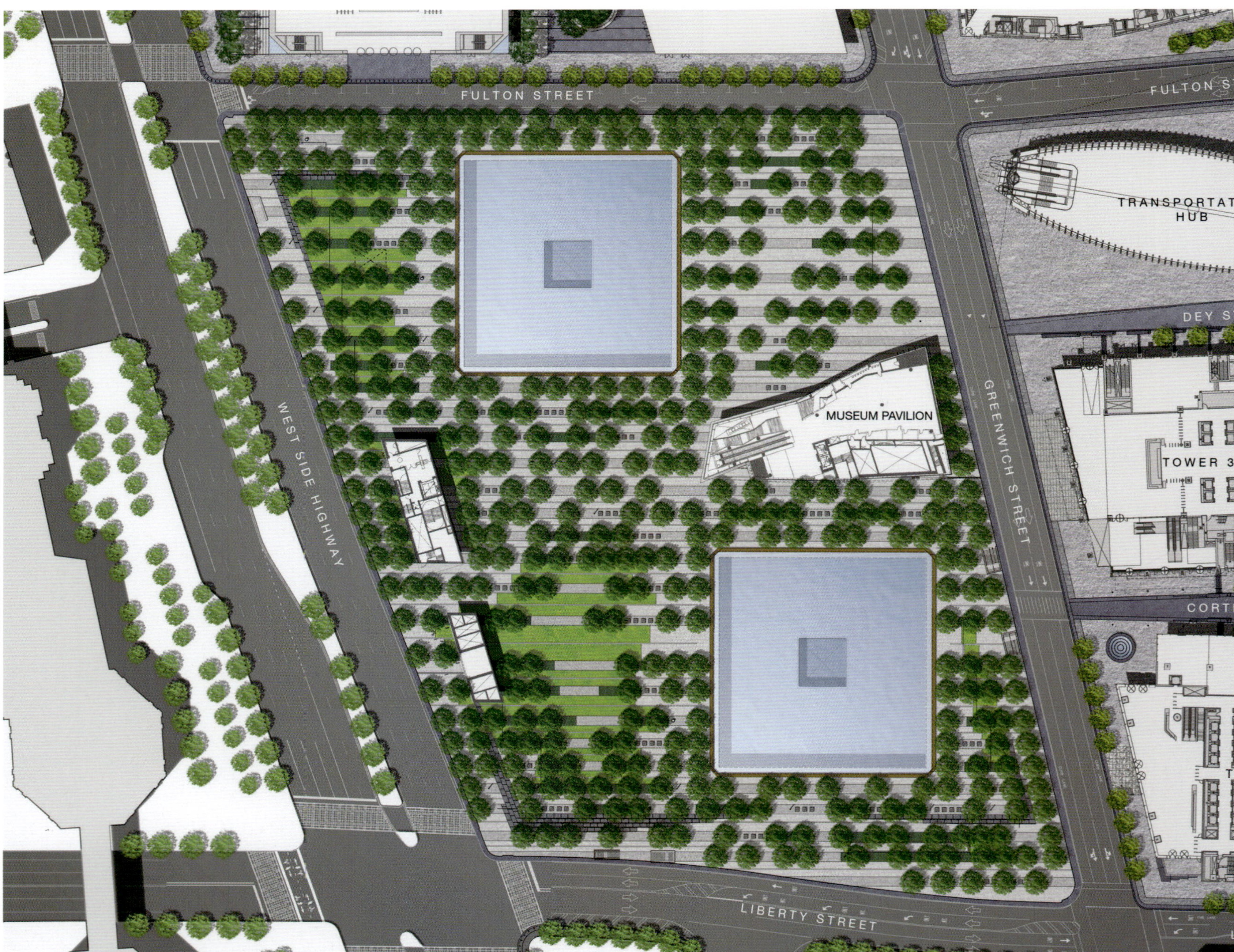

· **Photographer** - PWP, Alan Ward , Squared Design Lab · **Client** - National September 11 Memorial and Museum

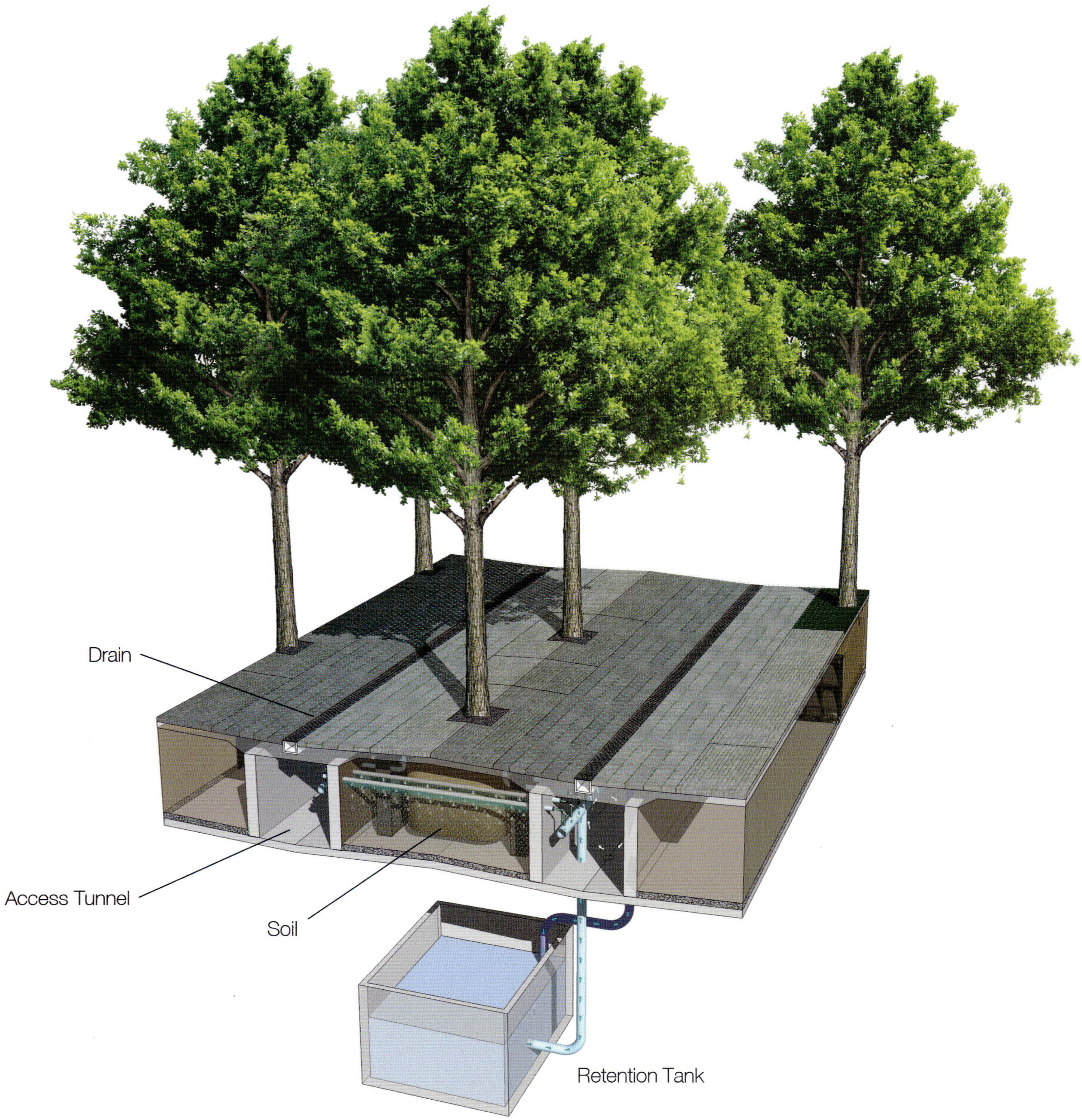

PWP 景观工作室，联合迈克尔·阿拉德（汉德建筑）以及戴维斯、布罗迪工作室，共同设计建造了一个纪念碑广场，缅怀 2001 年 9 月 11 日在世界贸易中心，宾夕法尼亚沙克斯维尔，五角大楼以及 1993 年 2 月 26 日在世界贸易中心受袭的遇难者们。设计师在原来双子塔的废墟上，设计了两块向下凹陷的喷泉景观，而在纽约城中心新的世界贸易中心大楼的周围，种满了橡树，为人们在这个新生的城市中心，提供了一个静谧安宁的可以沉思和回忆的地方。

纪念碑外形质朴简单，绿色顶部十分厚重，是一座完全符合生态学的建筑，内部是一个多功能的建筑空间，包括 PATH 公司地铁站台和轨道、纪念碑博物馆、中央制冷站、停车场以及一些辅助的基础设施。游客们在这里可以远离日常工作的喧嚣，徒步走进这个 416 棵橡树构成的神圣的地方。夏天，橡树叶在高高的上方形成穹顶，为人们提供阴凉；秋天，红黄间杂的树叶，又体现了季节的变换；冬天，阳光照在光秃秃的枝条上，在地上投射出斑驳的投影；春天，树木吐芽，大地重生。周而复始，生生不息，惨痛的 9.11 灾难在这里被赋予了永恒的概念。喷泉占地约 4047 平方米，向下凹陷深达 9 米多，周边是类似瀑布的流水。设计师们建造了全尺寸的模型，研究水流的效果，最终选择设计了节水节能效果最好的圆锥形堰坝，同时获得了最好的视觉效果。入夜，在灯光的映衬下，瀑布依旧静静流淌，波澜不惊。

在整个设计规划和建筑施工过程中，无论从材料的坚固性，还是景观的视觉效果角度，设计师们都优先考虑了广场的维护和持久性。广场地面和排水设施设计成一个自给自足的巨大蓄水系统。降雨和雪融化后形成的水经过管道排入巨大的蓄水池中，经过处理，重新利用，通过一套特殊的滴水和喷洒系统为广场树林提供灌溉。

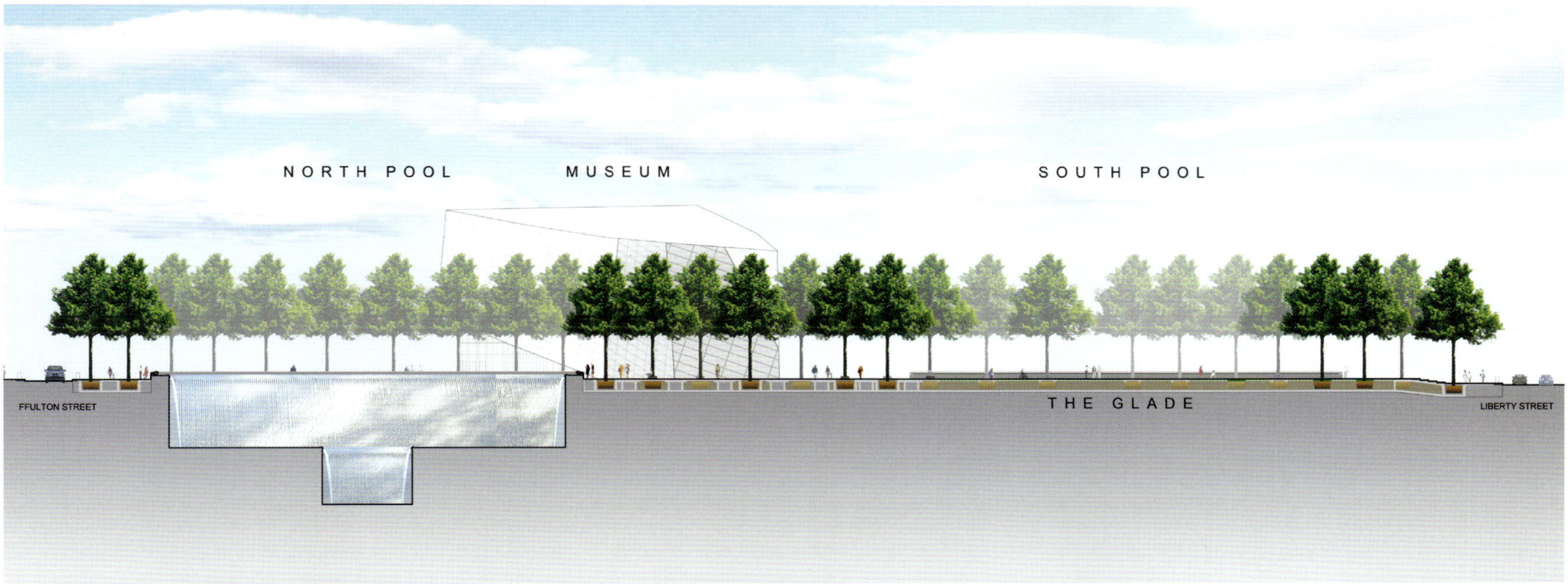
NORTH POOL
MUSEUM
SOUTH POOL
FFULTON STREET
THE GLADE
LIBERTY STREET

CHARLES J. HOUSTON
JOHN J. KREN
WILLIAM A. MATHESEN

• **Landscape Architect** - Enric Batlle, Joan Roig • **Firm** - Batlle i Roig Arquitectes

OCELLETS PARK

Barcelona, Spain

奥斯莱特公园

The rearrangement project for the "Ocellets" Park (Little Birds Park) intends to recover the site as a place devoted to urban leisure and entertainment, as well as to facilitate pedestrian traffic in all directions, especially towards "La Florida" primary care clinic.

The proposed solution aims to create a new landmark within the districts of Pubilla Cases and La Florida, therefore obtaining a new area to carry out diverse popular events, such as Castellers. This generates an open-plan area with enough capacity for a great number of people, and a sequence of surrounding zones that can be equipped for different applications and where people can sit on tiers with different dispositions. These tiers, apart from offering these areas, help us to cope easily with the elevation difference between the central square and the perimeter streets.

The central section is located on a practically horizontal plan, with a constant longitudinal gradient of 1.5% and transversal gradient of 1%. This enables surface waters to run under gravity, thus avoiding the formation of puddles and excessive run-offs. The rest of the areas, such as the "pipican" (dog waste disposal), the children's playground and the outdoor fitness zone, are also located on sloping planes with variable gradients (2.8% in the worst case).

The park's central square and the rest of the areas are paved with drainage materials: coarse sand in the central square, perimeter areas and fitness zone; and gravel of diverse grain size in the "pipican" and playground areas.

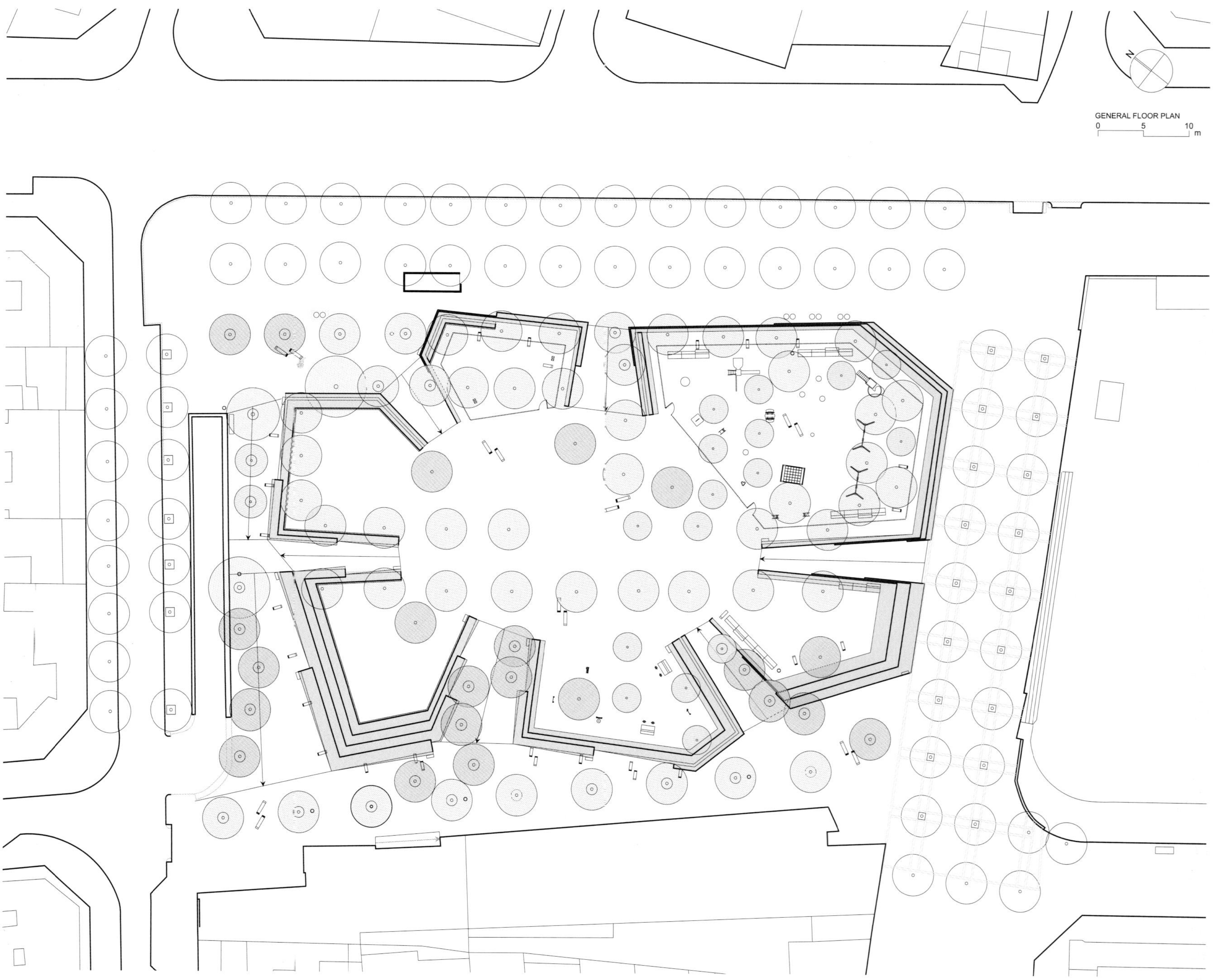

• **Collaborator** - Iván Sánchez (architect); Javier Zaldívar (architect & landscape architect); Dolors Feu (agriculturist & landscape architect); Elisabeth Torregrosa (building engineer); Lourdes Romeo (architect)

• **Photographer** - Jordi Surroca

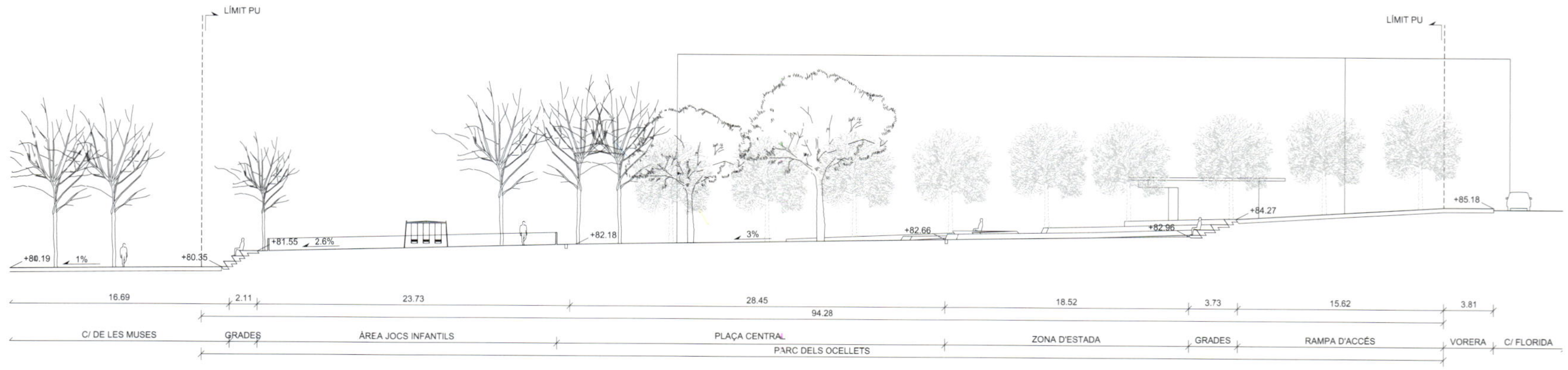
LÍMIT PU
LÍMIT PU
+80.19
1%
+80.35
+81.55
2.6%
+82.18
3%
+82.66
+82.96
+84.27
+85.18
16.69
2.11
23.73
28.45
94.28
18.52
3.73
15.62
3.81
C/ DE LES MUSES
GRADES
ÀREA JOCS INFANTILS
PLAÇA CENTRAL
PARC DELS OCELLETS
ZONA D'ESTADA
GRADES
RAMPA D'ACCÉS
VORERA
C/ FLORIDA

SECTION ST1

Taller Ba

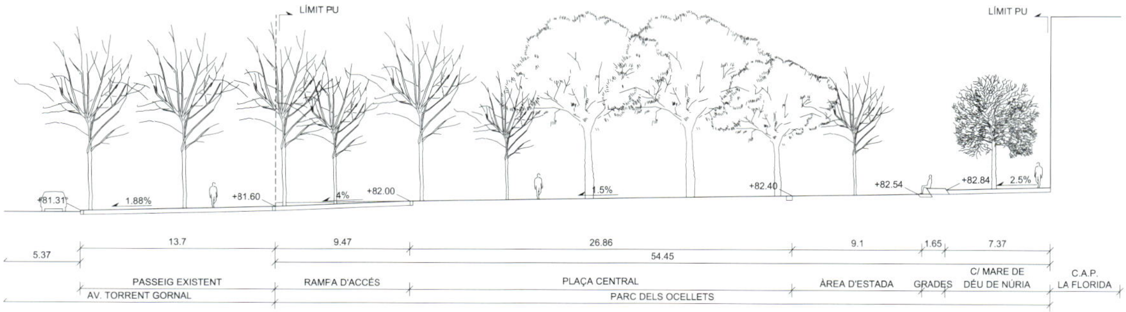
LÍMIT PU
LÍMIT PU
+81.31
1.88%
+81.60
4%
+82.00
1.5%
+82.40
+82.54
+82.84
2.5%
5.37
13.7
9.47
26.86
54.45
9.1
1.65
7.37
PASSEIG EXISTENT
AV. TORRENT GORNAL
RAMPA D'ACCÉS
PLAÇA CENTRAL
PARC DELS OCELLETS
ÀREA D'ESTADA
GRADES
C/ MARE DE DÉU DE NÚRIA
C.A.P. LA FLORIDA

SECTION ST2

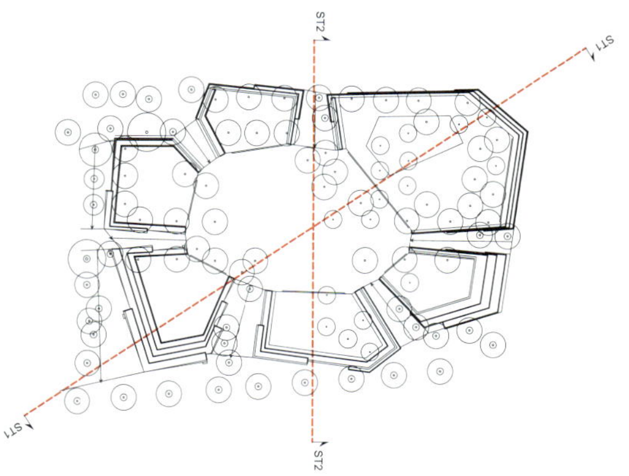
ST2
ST1
ST1
ST2

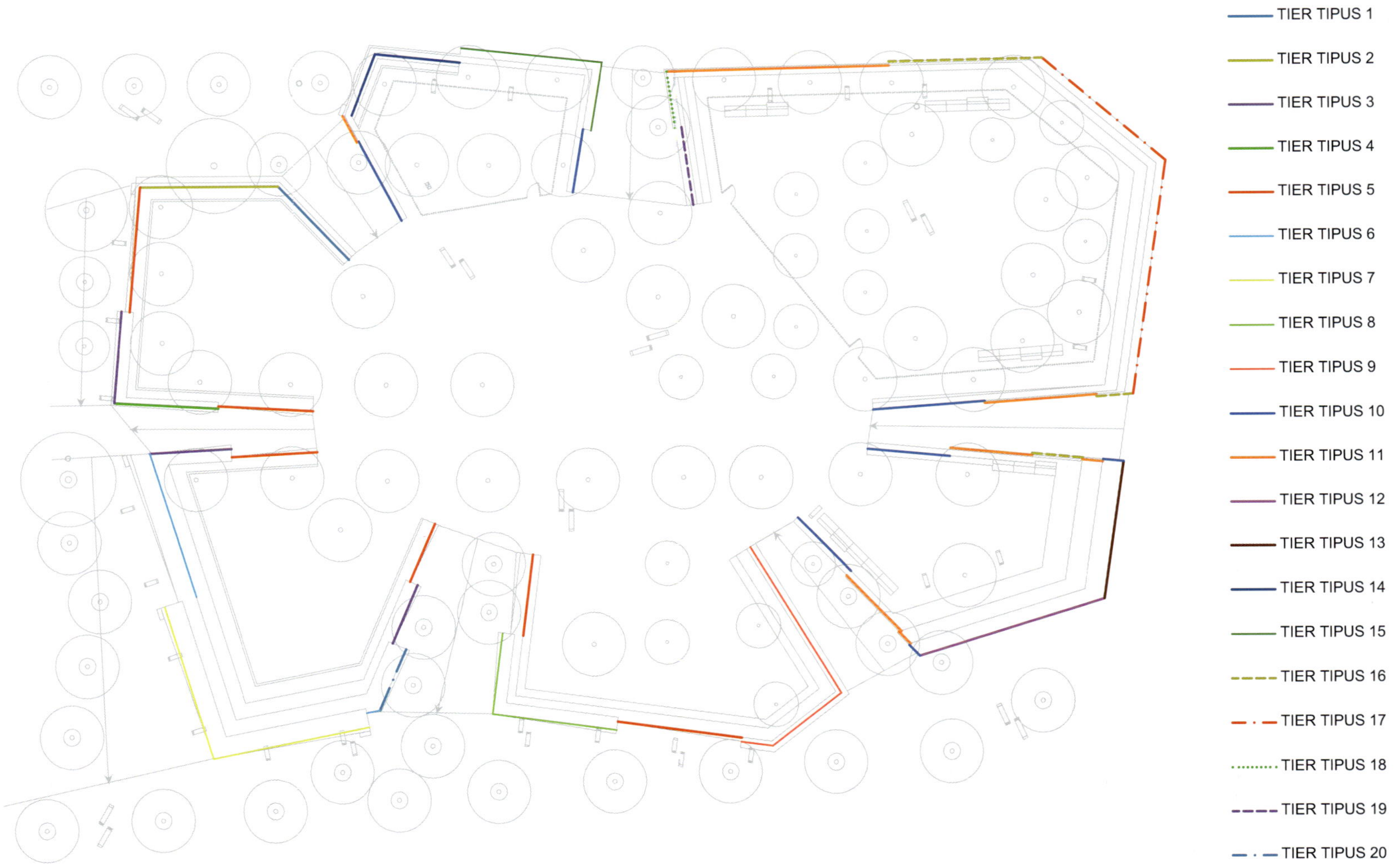

CATALOGUE OF TIERS

0 5 10 m

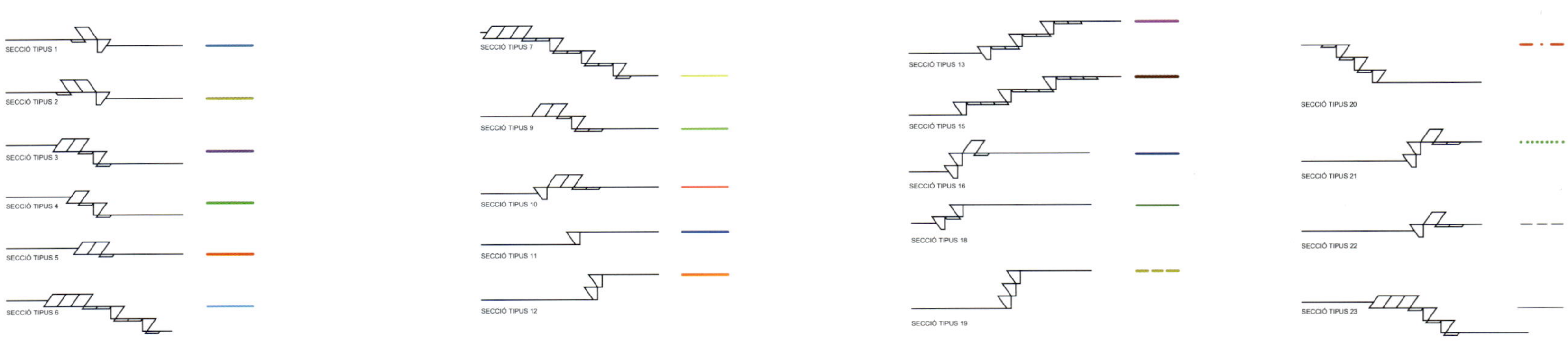

SECTIONS OF DIFFERENT TIERS

0 1 5 m

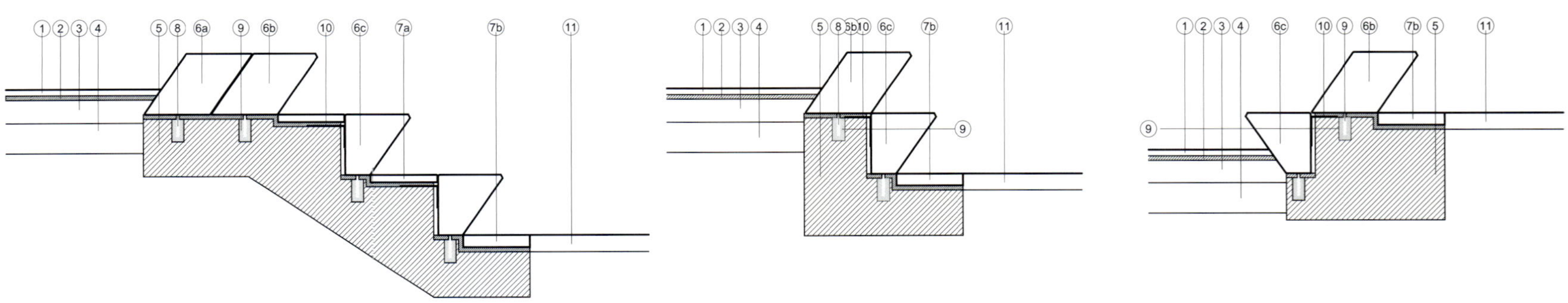

DETAIL OF TYPICAL TIER

0 0.5 1 m

1. Paving of grey, four-piece tiles
2. Layer of mortar, t = 3 cm
3. Concrete base, t = 15 cm
4. Sub-base of ballast, t = 20 cm
5. Reinforced concrete base
6a. Prefabricated concrete piece for bench model 108, without bevel (Santa & Cole)
6b. Prefabricated concrete piece for bench model 108, with bevel (Santa & Cole)
6c. Prefabricated lightweight concrete piece for bench model 108 with bevel (Santa & Cole)
7a. Prefabricated concrete piece, 44.5x50.5 cm, t = 5 cm (same finish as bench model 108)
7b. Prefabricated concrete piece, 44.5x50.5 cm, t = 8 cm (same finish as bench model 108)
8. Tie bolt
9. Fast-setting cement
10. Tie strip
11. Granite sand, t = 10 cm

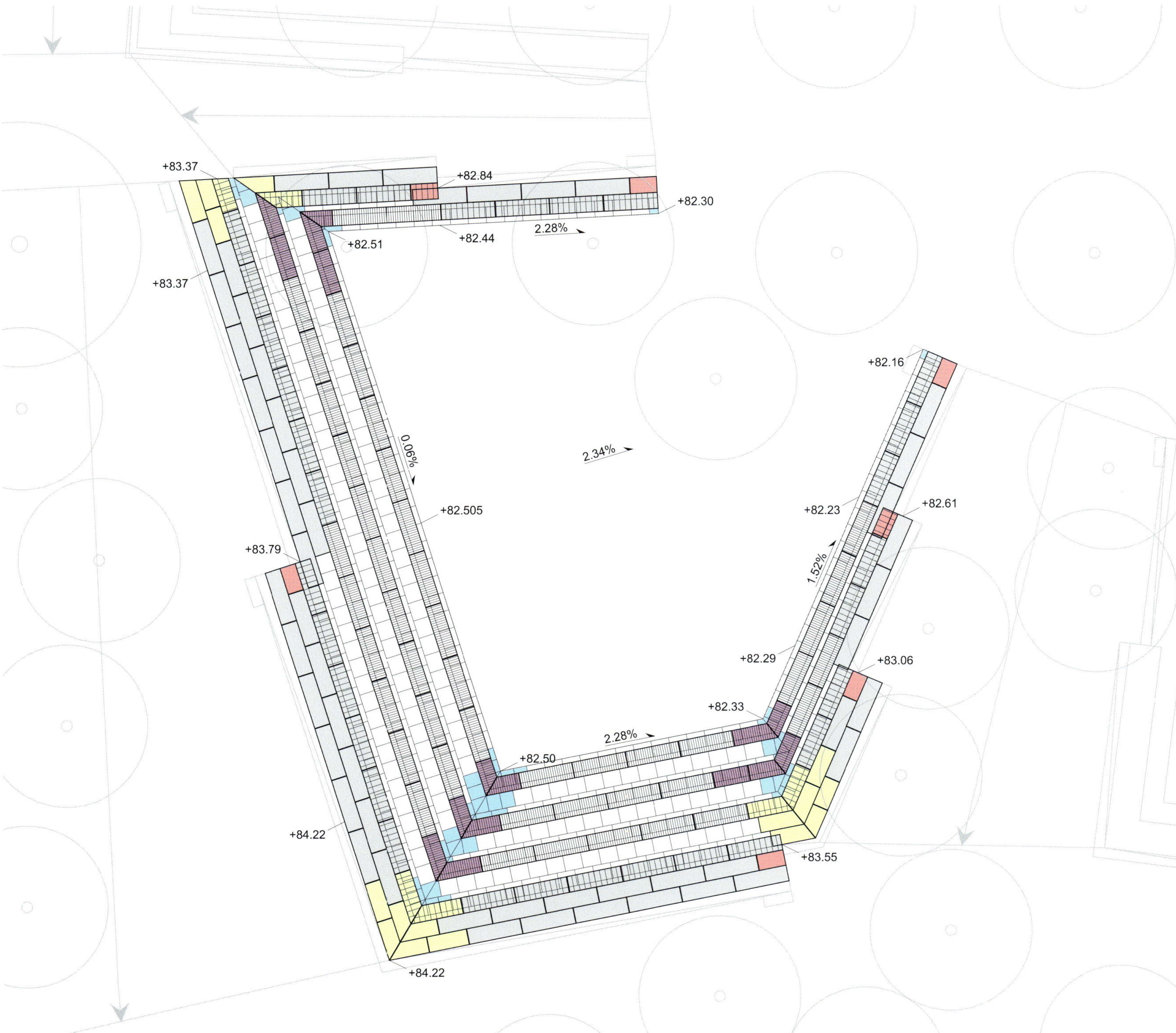

PERIMETRIC AREA BREAKDOWN. ZONE 2

0 1 2.5 m

BANCO 108 (ORIGINAL)	PLANTA	SECCIÓN
1.- ENTERO	150 cm	44 cm
2.- BISELADO	150 cm	38 cm

BANCO 108 (MITAD)	PLANTA	SECCIÓN
1.- ENTERO	75 cm	44 cm
2.- BISELADO	75 cm	38 cm

BANC 108 (ESPECIAL)	PLANTA	SECCIÓN
1.- ENTERO	variable	44 cm
2.- BISELADO	variable	38 cm

BANCO 108 ALIGERADO	PLANTA	SECCIÓN
1.- ENTERO	150 cm	44 cm
2.- MITAD	75 cm	44 cm
3.- ESPECIAL	variable	44 cm

PIEZAS AVANCÉ	PLANTA	SECCIÓN
1.- ENTERO	50,5 cm	44 cm
2.- ESPECIAL	variable	44 cm

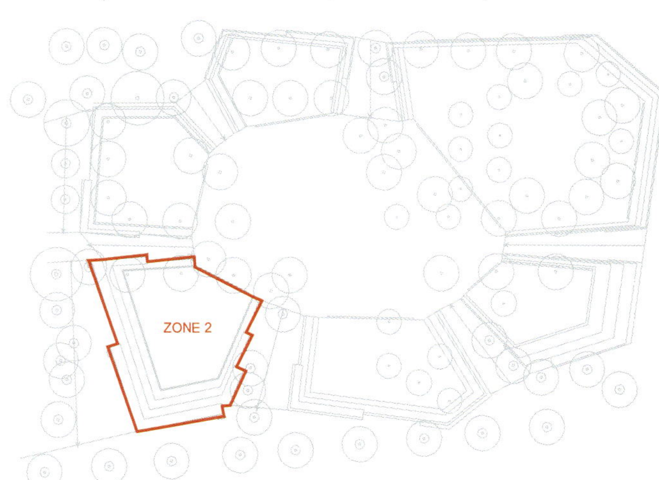

ZONE 2. GENERAL FLOOR

奥斯莱特公园（小鸟公园）重建计划的目的是通过重建，为社区提供一个休闲娱乐的公共场所，同时规划出四通八达的步行通道通向周边地区，特别值得一提的是，将修建一条专门通向佛罗里达初级保健诊所的步行通道。

设计方案的主要目的是在普比拉和佛罗里达社区打造一个新的标志性公园，为不同的公共活动提供场所，例如叠人塔活动。方案采取了开放式的平面布局，可以同时容纳大量民众进行活动，同时带动周边地区根据不同需要配置一些必要的公共设施，如人们可以在不同地段沿着成排层叠的台阶坐下小憩。这些成排层叠的设计距离中心地区较远，便于人们在中心广场与周边街道之间穿行。

广场中心部分十分平整，纵向坡度约为1.5%，横向坡度仅为1%。雨水流经地表，受重力影响，自然流向低处，避免了地面积水以及短时间内形成较大的水流。广场其他区域包括废物处理站、儿童游乐场、户外健身区等等，都位于不同坡度的坡面上（最大坡度为2.8%）。

公园中心广场以及其他平整的地面都由排水材料铺成：中心广场、广场周边以及户外健身区使用粗砂铺设地面；废物处理站以及游乐场则由不同粒径的沙砾铺设而成。

THE TOWN HALL SQUARE

The Netherlands

市政厅广场

The Town Hall Square in Arnhem is renovated and expanded by public space designer Ruud-Jan Kokke.

The square and court are paved with a pattern of black and white concrete tiles with an anti dirt top layer. The square is expanded in front of the entrance to give it more expression. Also the platform and steps are enlarged. A stainless steel ramp is constructed for wheels and elderly people. It all highlights the entrance of the City Hall. Kokke also designed the street furniture such as planters, benches and bicycle clips. In the stainless steel runner is written the history of Arnhem.

阿纳姆市市政厅广场由公共建筑设计师路德·简·科克进行翻新和扩建。

广场以及周围的地面铺满了表面防污的黑白相间的混凝土地砖。广场从市政厅入口处向外延伸，使市政厅看上去更加壮观。同时，平台和台阶也扩大了，还有为老人和残障人士修建的不锈钢坡道。所有的设计都用来突出市政厅的入口。科克还设计了街道上的其他装饰物，如花盆、长凳和自行车架。阿纳姆市的历史印刻在不锈钢步道上。

MATHILDE SQUARE

马希尔德广场

Eindhoven, The Netherlands

At the foot of one of Eindhoven's most important industrial icons, the Philips Light Tower, Buro Lubbers designed a new square. A pergola overgrown by wisteria and roses offers passersby a transparent glimpse of the tranquil and green atmosphere. Here one can, after shopping or a long day at the office, quietly enjoy a drink on a terrace or sit down on a bench among lush plants to escape the hustle and bustle of downtown Eindhoven.

The amorphous form of the planning area demanded an exact structure that brings serenity and also guarantees an optimal view of the Light Tower. The design concept is therefore based on rigid lines that run counter to the building. A feel of human scale is the result. The unity of the design is kept by using consistent shapes and materials: a dark gray concrete slab, robust Corten steel planters, and warm looking wooden benches. The gray carpet is laid in a rigid, complex pattern and is surrounded by a plinth that distinguishes the square from the building and also highlights the difference in level between the parking deck and the environment.

Elongated Corten steel planters create a stripe pattern. This produces alternating open and closed spaces, suitable for routing and terraces. The stripes of varying length, width and height integrate wooden benches and bicycle stands.

Mathilde Square was a challenge not only in terms of its design and urban integration requirements, but also in its demand for technical finesse. Located on the concrete deck of a parking garage, there was little space between deck and ground level for drainage, the paving and other engineering structures. Furthermore, the square had to be kept clear for emergency services. Despite these conditions an ingenious drainage system of labyrinth drip hoses and a keen layout of planters and tiles are realised which add up to the overall quality of the square.

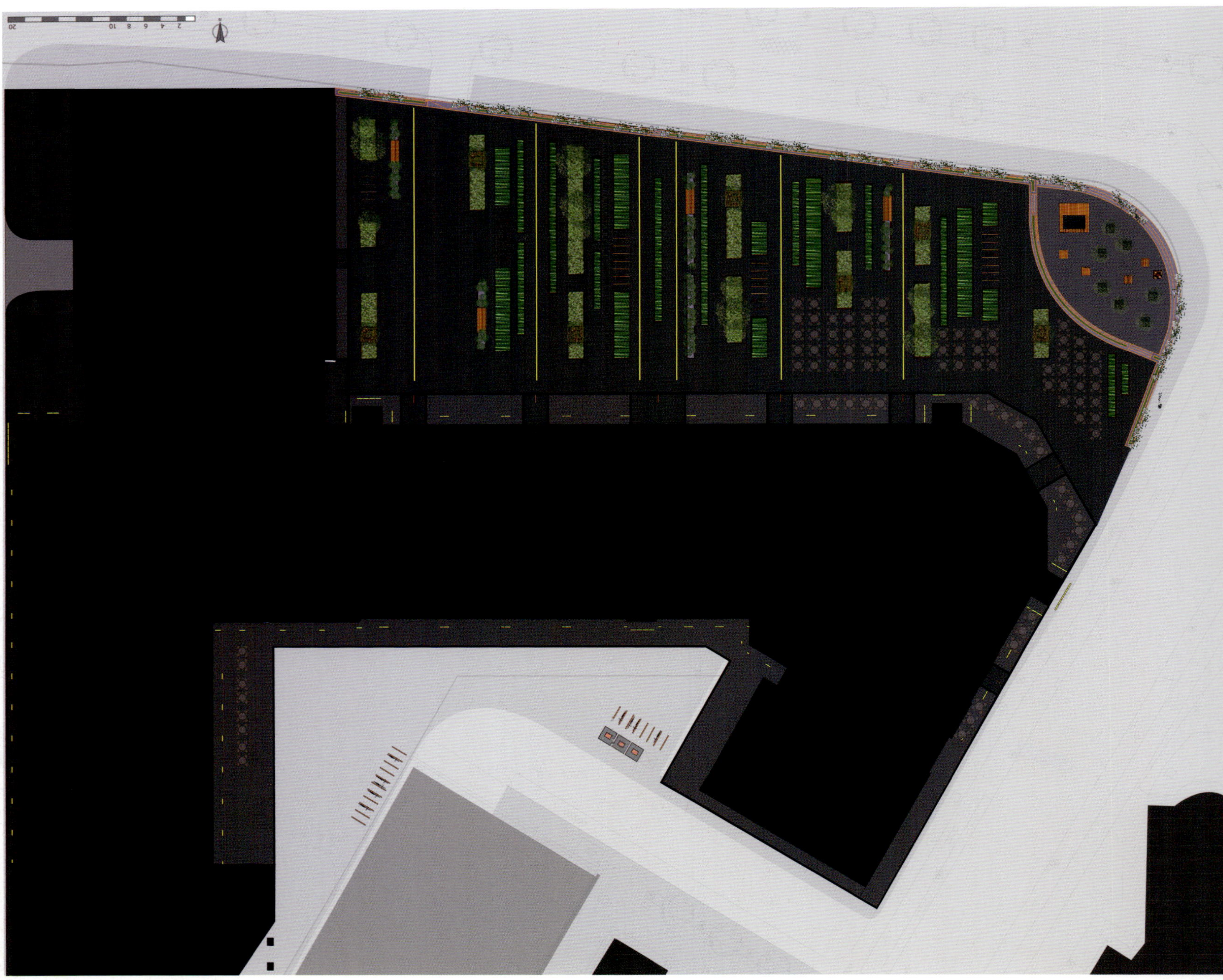

• **Area** - 5,500 m^2 • **Photographer** - Buro Lubbers

early spring

spring

summer

autumn

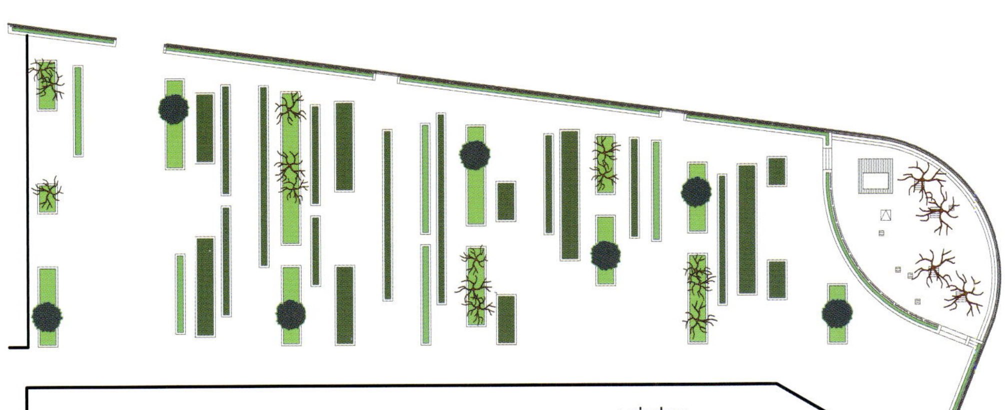

winter

SECTION A-A

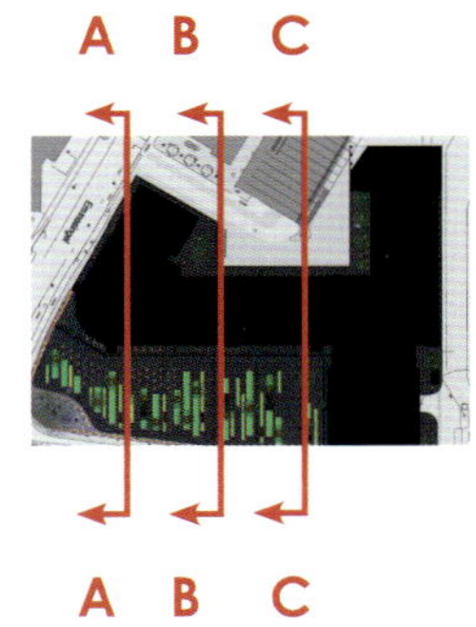

SECTION B-B

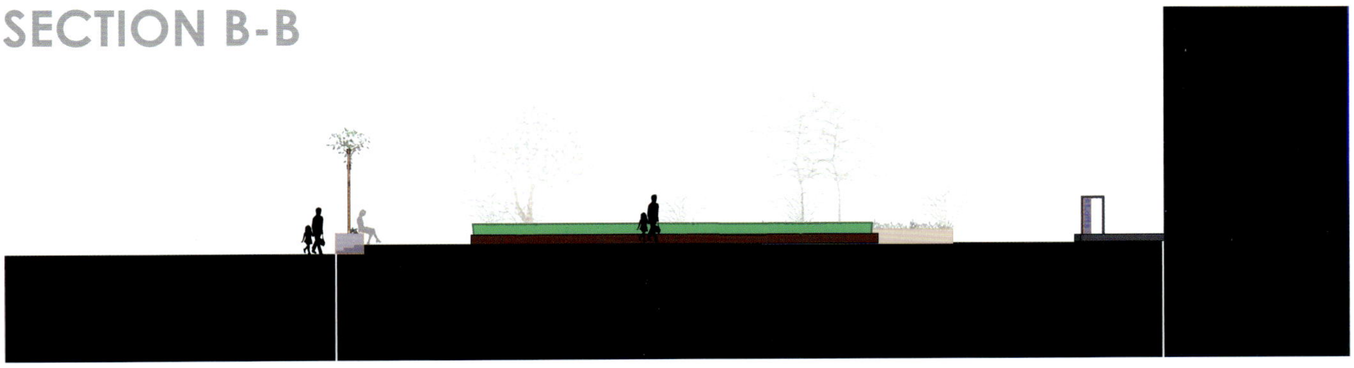

SECTION C-C

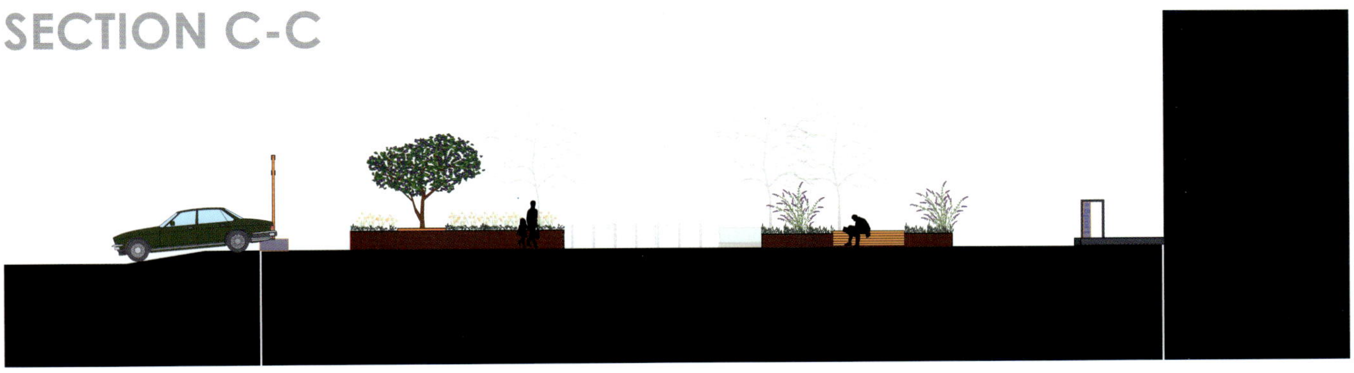

SECTION D-D

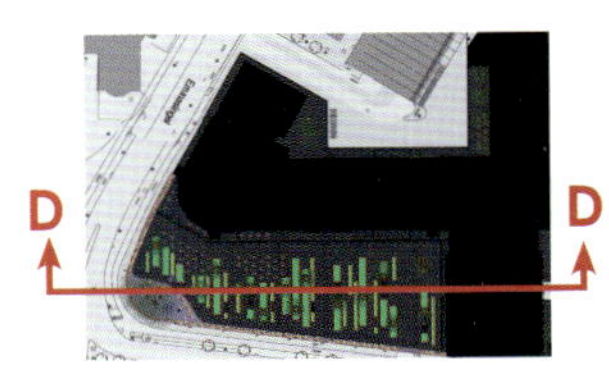

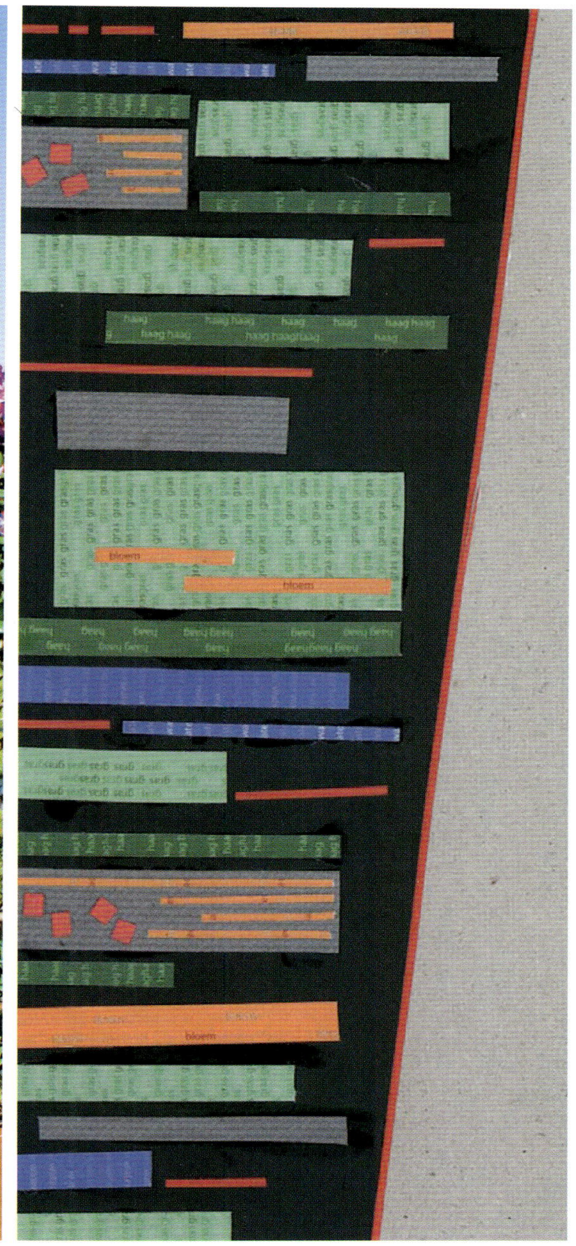

A

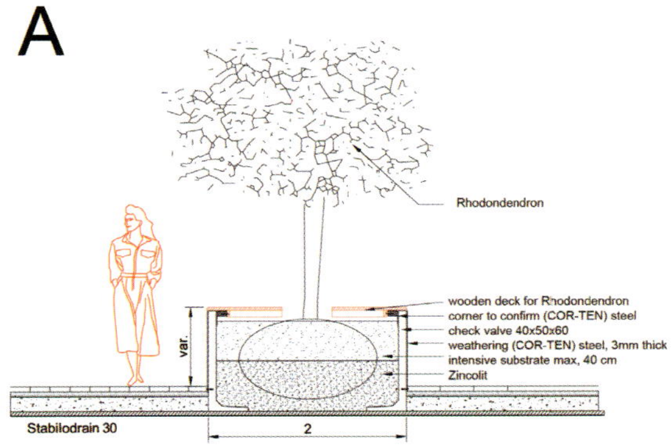

Planting stripe A with trees and perennial plants

GREEN STRUCTURE

The green structure of the square is formed by the rising structure of single trees and at ground level by a range of green strips with plants varying in kind. The planting is based on the underground car park, requirements for each type of plant are located in the strips. The height, length and width of these plants vary. The diversity of strips and space provides an opportunity for routing, shelter and terraces. The strips will be attractive by the use of evergreen plants and seasonal flowers.

B

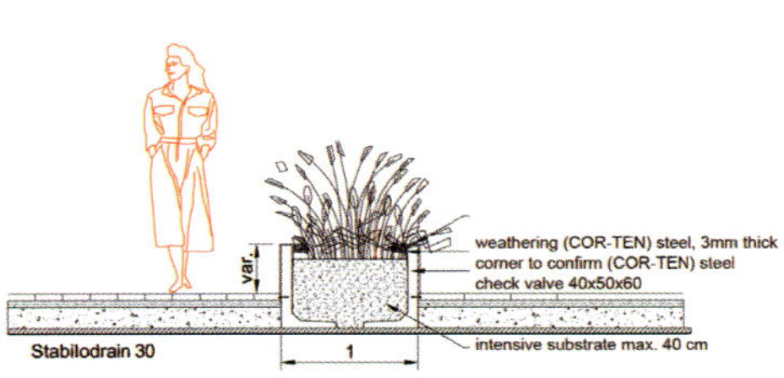

Planting stripe B with hedge

C

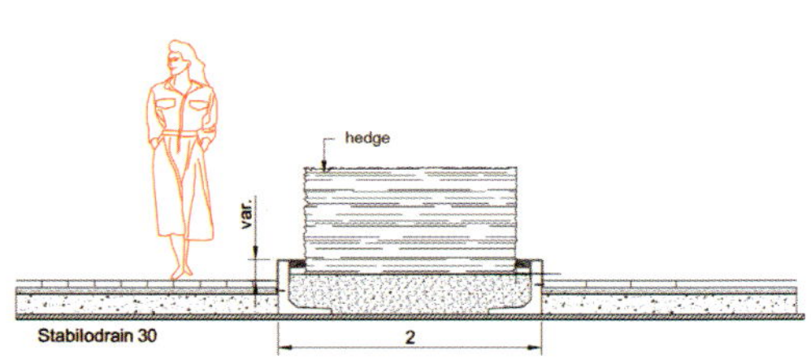

Planting stripe C with hedge

D

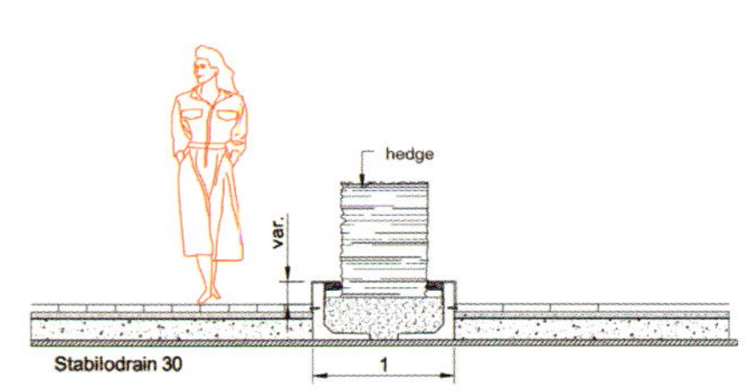

Planting stripe D with hedge

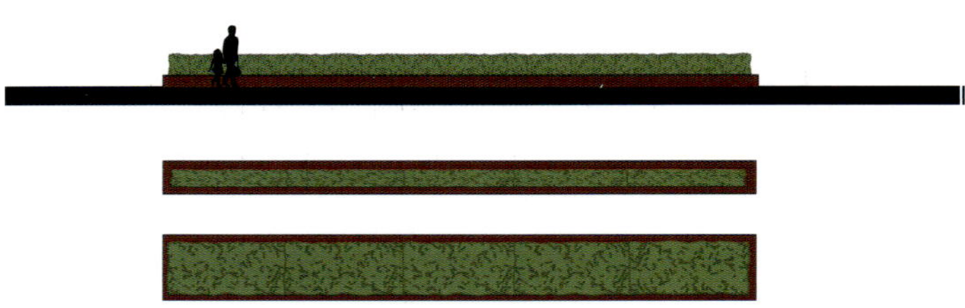

Hedge strip of variable width

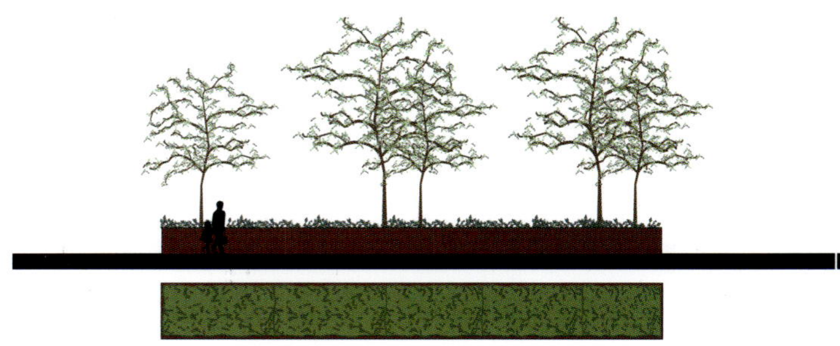

Tree clumps in strip

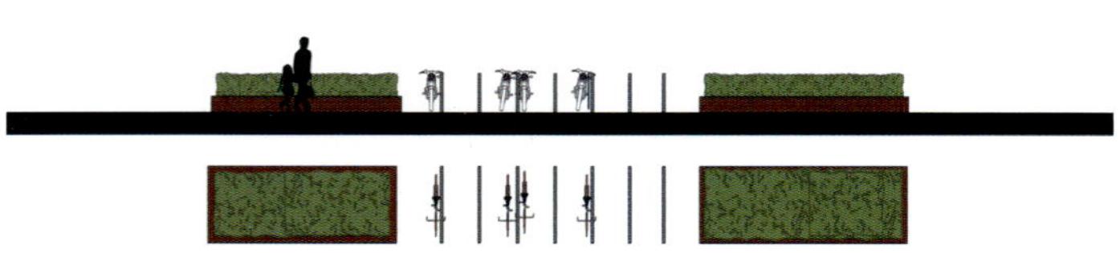

Bicycle parking included in strip

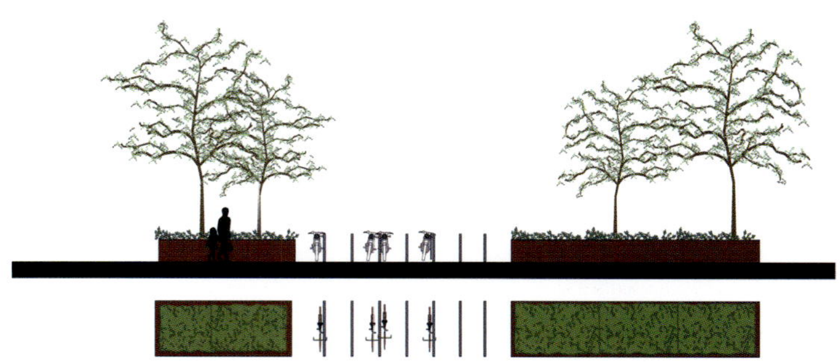

Bicycle parking included in tree strip

Bench included in strip with perennial plants

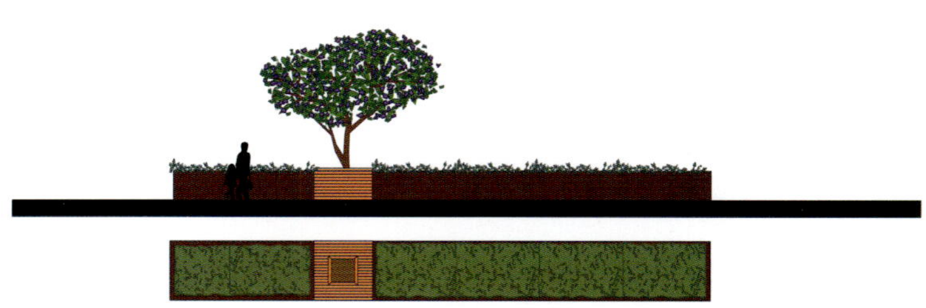

Special solitary tree with a wooden deck

菲利普灯塔是埃因霍温市最重要的工业地标之一，由布罗·鲁伯斯事务所设计完成的马希尔德广场就位于灯塔脚下。广场上安静祥和的小路上方，柴藤和玫瑰属植物爬满了藤架，路人通过藤蔓的空隙可以看到别样的风景。在一天的辛勤劳作或购物后，人们可以在这里的天台或长椅上安静地享用一杯醇香的咖啡，或是远离喧嚣和忙碌，在郁郁葱葱的草地边享受埃因霍温傍晚的静谧。

设计师们为整个广场的无序形状制定了严格的结构规划，营造出一种宁静祥和的氛围，同时，保留了观赏灯塔的最佳视角。设计的主要理念是大幅运用与建筑本身形成强烈反差的粗硬线条，一种强烈的人性深处的震撼感扑面而来。设计师们利用一系列的建筑形状和材质来综合表现整体设计风格：暗灰色混凝土路面、粗犷的柯尔顿钢制绿化架和暖色调的实木长椅。缓冲带上铺有硬材质的灰色复合地毯，四周是突起的底座，不仅表明了广场和建筑之间的界线，同时，也突出了停车场顶棚与地面之间的水平差异。

柯尔顿钢制绿化架呈条纹状向外延伸，隔离出开放空间和闭合空间，这样可以在适当的位置设置通道和台阶。不同长度、宽度和高度的条纹形状顺势分布，与实木长椅和自行车停放处融为一体。

马希尔德广场的设计规划十分具有挑战性，这种挑战性不仅来自于项目本身的设计和城市一体化进程的需要，更重要的是来自于技术处理方面的要求。广场位于一个地下停车场的上方，停车场顶棚与地面之间空间狭窄，很难满足排水系统、路面摊铺以及其他工程结构的需要。另外，广场上要保持空旷，以备应急服务部门的临时征用。尽管施工条件有限，设计师们还是在保证整个广场工程质量的基础上，巧妙设计并完成了迷宫般的排水系统和园艺陈设体系。

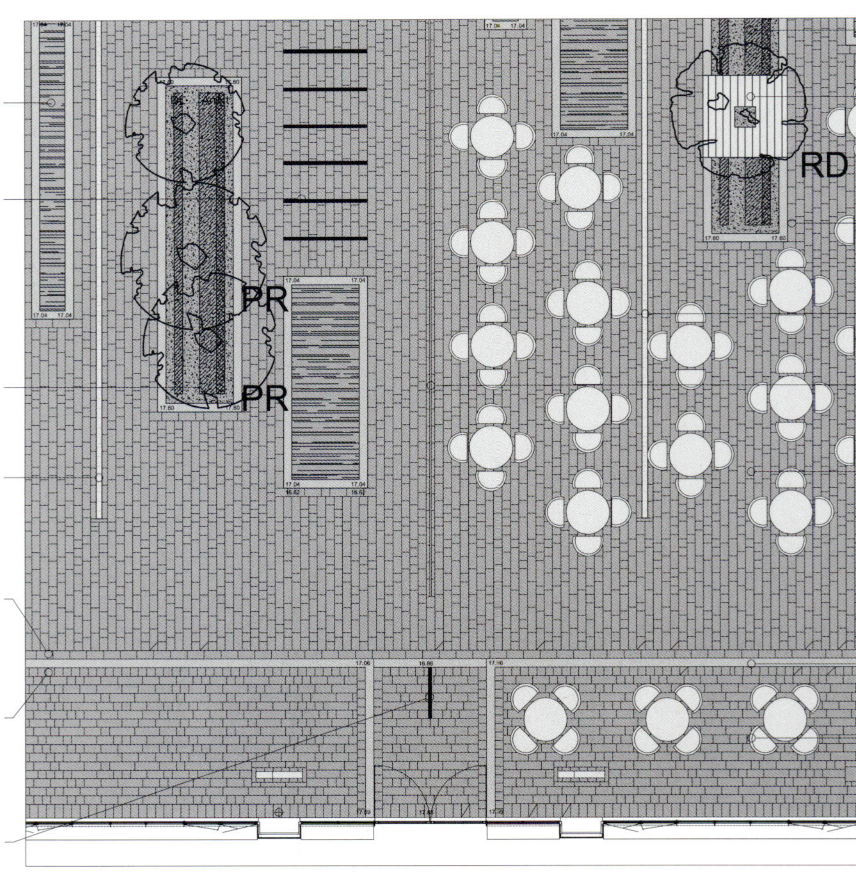
Planting strip 1m wide:
hedge: ilex meserveae "Blue Prince"
Bicycle parking: length 200cm, long steel L profile (8cm), supplier: VCP streetcare
Planting strip 2m width: groundcover: pachysandra terminalis with bulbs (see planting plan) in strip 3x prunus yedoensis size 20-25
Gutter line: ACO multiline V100 with cast iron edge Tiles along gutter line cut to size
Brickwork: A'gora tiles polished without auction. Color bluestone. Supplier. Bleijko. Size: 200x200x80mm
Brickwork: A'gora tiles polished without auction. Color bluestone. Supplier. Bleijko. Size: 200x200x80mm
Bicycle parking with mounting options for advertising: length 120cm, long steel L profile (8cm), supplier: VCP streetcare
PR
PR
RD
Planting strip 2m width: groundcover: pachysandra terminalis with bulbs (see planting plan) wooden deck with planting container: with Rhodondendron "Catawbiense Boursault"
Brickwork around planting container: A'gora tiles polished without auction color Bluestone. Supplier Bleijko. Size 200x400x80mm
Gutter line: ACO multiline V100 with cast iron edge. Tiles along gutter line cut to size
Leading lights: Philips lightline. Tiles along gutter line cut to size
Pavement: A'gora polished without auction color Bluestone. Supplier Bleijko. Size 200x400x80 mm (20%) 200x400x80 mm (50%) 200x200x80 mm (30%)
Curb: 200x300x1000 anthracite
Pavement plinth: A'gora polished without auction color Bluestone. Supplier Bleijko. Size 200x400x80 mm (20%) 200x400x80 mm (50%) 200x200x80 mm (30%)

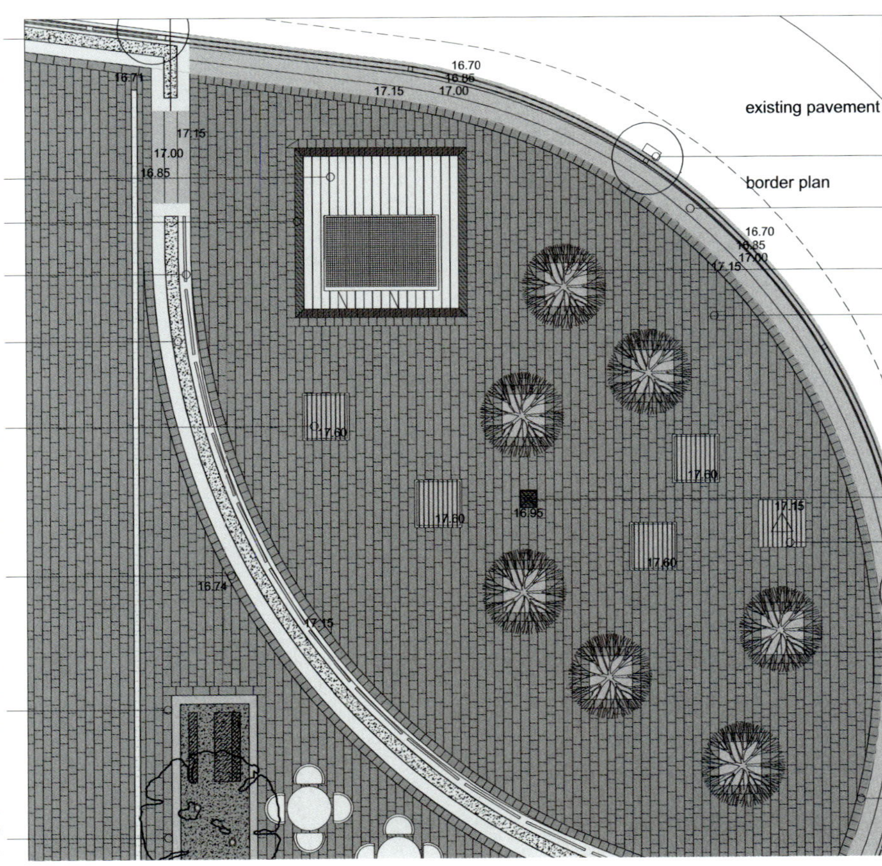
Concrete edge with color range. Color: dark gray with perennials
Conversion of wood to discharge garage
Drainage channel
Fence: panels
Concrete edge with perennials
Sample: wood durability class 1 with FSC certification
Brickwork: A'gora tiles polished without auction color Bluestone. Supplier Bleijko. Size 200x200x80mm
Brickwork: A'gora tiles polished without auction color Bluestone. Supplier Bleijko. Size 200x400x80mm
Brickwork: A'gora tiles polished without auction color Bluestone.
existing pavement
border plan
Saving in pavement for Wisteria
Stairs: Concrete with color range
Tree grate with Cedrus libani "glauca fastigiati"
Pavement: A'gora polished without auction color Bluestone. Supplier Bleijko. Size 200x400x80 mm (20%) 200x400x80 mm (50%) 200x200x80 mm (30%)
Drain: 200x200mm Connect to existing sewer.
Wooden base with "lampenmaakstertje" (public statue from the municipality Eindhoven)
Brickwork: A'gora tiles polished without auction color Bluestone. Supplier Bleijko. Size 200x200x80mm
Brickwork: A'gora tiles polished without auction color Bluestone.

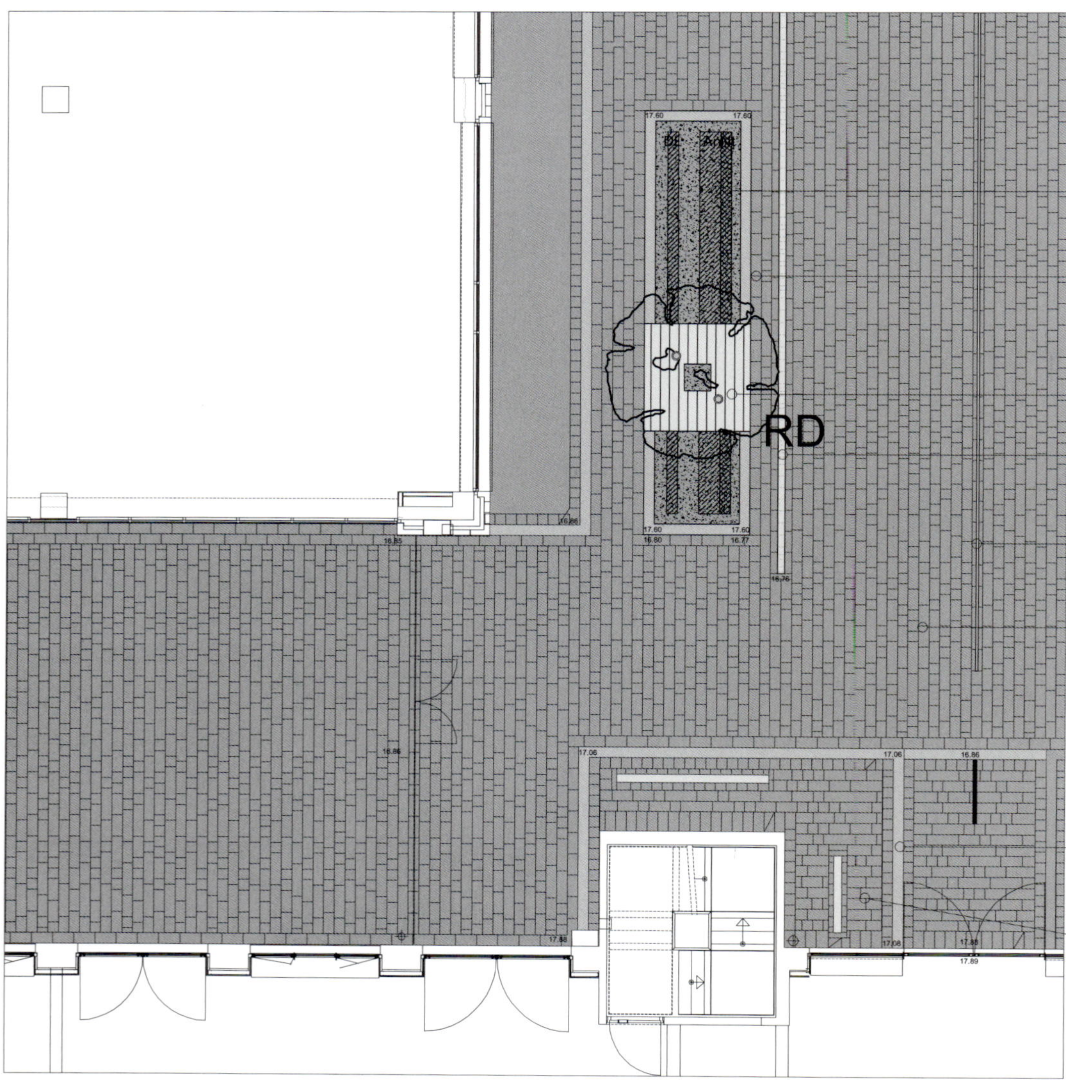
RD
Planting strip 2m width:
groundcover: pachysandra terminalis
with bulbs (see planting plan)
Brickwork around planting container:
A'gora tiles polished without auction color Bluestone.
Supplier Bleijko.
Size 200x400x80mm
wooden deck with planting container: with Rhodondendron
"Catawbiense Boursault"
Gutter line: ACO multiline V100 with cast iron edge.
Tiles along gutter line cut to size
Leading lights: Philips lightline.
Tiles along gutter line cut to size
Pavement: A'gora polished without auction color Bluestone.
Supplier Bleijko.
Size 200x400x80 mm (20%)
200x400x80 mm (50%)
200x200x80 mm (30%)
Curb:
200x300x1000 anthracite
Pavement plinth: A'gora polished without auction color
Bluestone. Supplier Bleijko.
Size 200x400x80 mm (20%)
200x400x80 mm (50%)
200x200x80 mm (30%)

• **Landscape Architect** - Gustafson Porter • **Client** - Royal Borough of Greenwich • **Area** - 10,000 m^2

WOOLWICH SQUARES

伍尔维奇的广场

Woolwich, England

The town center was fragmented due to the road layout which separated the green space of General Gordon Square from Greens End, Beresford Square and the rest of the town center. Gustafson Porter proposed a unified pedestrian space with two squares of different characters that complement and relate to one another, each providing an aspect of public life within the town. Greens End, a shared surface roadway linking the two squares, was designed with one material treatment and a single identity that forges it in people's mind as a real street rather than a sum of several parts.

In General Gordon Square, the quality of the garden is reinforced through the choice of trees and planting, with larger trees separating it from the surrounding roads. Lawn terraces step down gently to the water scrim which is fully accessible but can also be easily drained. Beresford Square is an altogether more functional space during the day informed by the daily market. In the evening the atmosphere changes with the use of catenary lighting to energize the square. The Royal Arsenal Gatehouse provides a backdrop for events and social activities after the market has left.

Terracing was designed in both squares to improve the usability of the steeply sloping squares, with level routes provided throughout to ensure accessibility for all users. A language of materials and details was developed that incorporated the best practice for seating, ramps, lighting and planting together with a fully accessible water feature. These features together with the protection of key sightlines through the squares were cited as reasons to award the project a Secure by Design award.

• **Photographer** - Gustafson Porter, Grant Smith, Chris Mansfield and Redshift Photography

ROYAL
GREENWICH

wimpy

ROYAL

NHS
CLOVER HEALTH CENTRE
GP Practice and Walk-in Centre

伍尔维奇市中心地带被道路交通布局划分为几个部分：绿草如茵的戈登将军广场、格林斯路口、巴利福德广场以及市中心区域的其他部分。广场项目的设计方案由古斯塔夫森·波特工作室负责，目的是合并现有的两个不同风格的广场，打造一个统一风格的步行空间；利用两个广场的设施和特点互相取长补短，为公众提供更好的城市公共生活空间。格林斯路口是连接两个广场的通道，虽然地面由一种材料铺设而成，但在公众的印象中，这里是一个街区，而不是几个区域拼凑在一起的。

在戈登将军广场，临街一侧栽种了大树，遮蔽噪音，大量栽种的不同树种大大提升了公园的品质。草地旁边的台阶坡度不大，直接延伸到雨水过滤系统中，通过这个系统可以迅速排出积水。巴利福德广场在白天是一个商品交易场所，是一个更具功能性的公共区域。入夜后，气氛完全不同，经常举行各种社交活动，远处矗立着皇家兵工厂的大门，悬链式的灯光系统也为广场平添了一丝活力。

为了更好地利用这个位于斜坡上的广场，在原有的两个广场上都设计规划了梯形看台，带有坡度的小路贯穿其中，方便行人通行。建筑材料和细节上的精心选择和设计，配合看台座椅、斜坡地面、灯光系统、绿化景观以及一个可以进入其中的水系景观，构成了一个具有当地文化气息的场所。凭借这些设计元素和通透的广场视野，广场项目荣获了建筑设计大奖。

INDEX

索 引

BJØRBEKK & LINDHEIM LANDSCAPE ARCHITECTS

Jostein Bjørbekk and Tone Lindheim founded Bjørbekk & Lindheim Landscape Architects in 1986. At present, the studio consists of 18 landscape architects and one landscape engineer. The firm today is one of Norway's largest and one of the leading Nordic firms within landscape architecture.

The essential design concept for all of Bjørbekk & Lindheim's projects is to pay attention to each individual locality. They do extensive research, and historical, social and cultural settings are incorporated into their projects. They look at how topography and climate, as well as natural vegetation and existing building traditions, have impacted on a place. Are there any stories, myths or legends that can be told, that can be dramatized? What are future needs that they should accommodate? How can the aesthetic and functional challenges be solved within the prevailing economic, cultural and natural limits? Bjørbekk & Lindheim Landscape Architects try to find simple answers to complex problems, designing the site so that it looks right, as if it has always been there.

They try to create beauty in everyday life, to make attractive meeting places and functional landscapes which will age with dignity and mature gracefully. The firm also focuses on ecological solutions.

CLAUDE CORMIER + ASSOCIÉS INC.

CLAUDE CORMIER + ASSOCIÉS INC. is an internationally recognized practice that extends far beyond the conventional realm of traditional landscape design to forge bridges between urban design, public art, and architecture. The projects of CLAUDE CORMIER + ASSOCIÉS INC. are likewise anything but conventional, celebrating the artificial and surreptitiously altering reality with the surreal one. The team work is distinguished not only for its inventiveness, but also for its tenacious optimism in the power of design.

Since 1995, the firm has been privileged to work on major public work in Montreal and Toronto – the restoration of Square Dorchester, Place Youville in Old Montreal, Lipstick Forest at the Palais des congrès de Montréal, and collaboration with the Old Port of Montreal since 2000. Also in Toronto – HtO Park and Sugar Beach on the Toronto Waterfront; the transformation and adaptive re-use of the Don Valley Brickworks into an ecological and cultural centre and the public landscape for the new Four-Seasons Hotel and Residences downtown Toronto. Cormier has taught at the University of Montreal, and has been invited to lecture across Canada, the United States, and countries in Europe. In 2009, Cormier was knighted with the Ordre National du Québec, the highest distinction bestowed by the government of Québec.

OMGEVING

OMGEVING is an independent, inter-professional group of landscape architects, urban planners, and architects. The group has set itself the social task of protecting the environment. As a result of this commitment, OMGEVING has been working both for the public sector and the private sector since 1973. During this time the group has become a medium-sized bureau with some 50 experts specialized in different areas.

PAREDES & PINO ARCHITECTS

Fernando Garcia Pino (1, Madrid, 1965) and Manuel García de Paredes (2, Madrid, 1966), graduated from Madrid School of Architecture (ETSAM), founded Paredes & Pino Architects in 2000.

Both have completed their training working on numerous projects, works and competitions, in renown offices. In the period from 1992 to 1997 Fernando Pino worked in the Mansilla and Tuñón office and later with Juan Navarro Baldeweg. Manuel García de Paredes worked from 1990 to 1999 in Paredes Pedrosa architects practice.

Since 1998, they have won several national and international awards of architecture, such as Europan 5 and 6, the Citizens Activity Open Center in Cordoba (CAAC), housing for the Municipal Housing and Land of Madrid (EMVS), the building of the Administrative Courts in the City of Justice of Madrid, the Olympic Gymnastics Hall and the Access and Services building to the Olympic Village for the 2016 candidate city or Yebes multifunctional center in Guadalajara.

Both have been guest lecturers in numerous national and international universities. Fernando Pino is a professor in the Architecture Project Department at the ETSAM since 2004 and now they are also professors in the University of Architecture IE at the campus of Segovia.

XPIRAL& MTM

XPIRAL and MTM have developed in the last 15 years projects of great scale in different places from optics of integration for infrastructures and architectures in an only social space, for diverse private and public clients in all Spain, from National and Regional Administrations, City Councils and Public Corporations.

They converge in their proposals landscape conditions, architectural, economic and technical aspects that are developed next to Spanish and European teams specialized, for example, The Technological Center of the Metal in Murcia in the year 1998; Singular building and Ordination of the Public Space Las Vegas, in Corvera of Asturias in the year 2000; City Rises in Córdoba in the year 2002; and the International Competitions EUROPAN V in Paternal, VI in Bilbao and VII in Burgos, obtaining magnificent fruit of which have come out different responsibilities.

ØSTENGEN & BERGO AS

Østengen & Bergo is a Norwegian landscape architectural company, consisting of landscape architects with different backgrounds and experiences. This richness makes them able to solve landscape challenges on most levels. They are working on a wide range of tasks within landscape architecture, from landscape analysis and regional planning to design of streets, parks, squares and plazas. Inspiring and functional designs are significant goals for all their work, as well as being based on characteristics and context. They promote sustainable, green and functional solutions, for interaction, inspiration and relaxation. The office is eagerly trying to develop a greener and lush landscape in urban situations. The company is well known for sober and simple design of parks, urban plazas and school areas. Østengen & Bergo has won several architectural competitions in the last years, and has received prestigious architectural prizes.

OLM

OLM, Office of Landscape Morphology, is a landscape firm established in Paris in 2005 by Philippe Coignet, French landscape architect. Philippe Coignet studied Geography at the University of Pantheon Sorbonne in Paris and at the Graduated School of Landscape Architecture in Versailles. He holds a Master's degree in Landscape Architecture from the University of Pennsylvania in Philadelphia. Philippe is also involved in teaching landscape architecture design in various universities, as the ETH Polytechnical Architecture School in Zurich and Ecole Nationale Supérieure d'Architecture in Paris Malaquais. Since 2010, Philippe has taught a workshop and studio at GSD in Harvard University, USA.

In 2008, OLM received the prestigious award of Best Young Landscape Architect from the Ministry of Culture.

OLM realizes numerous studies and missions for public consultants in France and abroad which concern the transformation of urban territories, transformations of airports, tourist complexes, urban parks, cultural gardens, urban landscape and requalification of industrial sites.

OLM works daily with numerous partners, such as ecologists, hydrologists, architects, town planners, light designers and technical engineering consulting firms.

ADEPT

ADEPT is based in Copenhagen, Denmark and works within the fields of architecture, urban planning and landscape design. ADEPT is founded and led by architects: Martin Laursen, Martin Krogh and Anders Lonka.

ADEPT is currently working with projects "Flintholm Spark" 4000 m^2 House of Culture and Movement at Frederiksberg (DK), "Dalarna Media Arena" 3600 m^2 multimedia library in Falun (SE), "The Spacial City" future proofing of the 55,000 m^2 housing complex Egedalsvæge (DK), and urban planning projects "The Tolerant City" 1,000,000 m^2 urban development in Helsingborg (SE), "Fredericia C" turning an industrial area into a new urban area as well as small-scale urban space projects: a square in Silkeborg (DK) and a pocket park in Copenhagen (DK).

ADEPT's three partners have all been engaged and involved in teaching, workshops and lectures since they founded the company. Anders Lonka teaches in sustainable urban planning at the School of Architecture at the Royal Danish Academy of Fine Arts, while Martin Laursen and Martin Krogh are external professors at the Institute of Architecture, Design and Media technology at AAlborg University.

GARTEN STUDIO LTD.

Garten Studio Ltd. has three owners, and usually works with six to eight employees, all of them are landscape architects. They usually do the landscape design part, but involving other skills they do masterplans too. Most of their jobs are from Budapest – capital of Hungary– and major Hungarian cities. They stand for contemporary landscape design, mostly in urban environment. They think that unique landscape objects, proper material usage play important part in making public spaces.

HOSPER

HOSPER is a design-office for landscape architecture, urban planning, outdoor space and object. The assignments which they work on are varied but always require the ability to generate new and creative ideas. They make up a young and enthusiastic team, within which several design disciplines are represented: landscape architecture, urban planning and industrial design.

Society continuously imposes different demands on its environment. Each new generation adds a new layer to the landscape. As designers, they attempt to translate the demands made by their times on the environment into a new layer. They retain respect for the layers which have gone before while creating new chances for the future. Their most important priority in doing this is to cherish the experience of space.

BUREAU B+B URBANISM AND LANDSCAPE ARCHITECTURE

Bureau B+B is an internationally operating design agency, founded in 1977. They believe in collective talent more than any other design agency. They rely on the shared intelligence of all their team members. Their designs are based on the unique aspects of a task, and they identify the elements a site offers. They then produce a functional design that always includes an inspiration or a disruptive component that allows the spatial experience to transcend the functional one.

ARQUITETOS ASSOCIADOS

Arquitetos associados is a collaborative studio dedicated to architecture and urban design based in Belo Horizonte, Brazil. Each project is treated as a unique and specific work, for what a particular organization of the group is set, allowing the emergence of various design groups inside the studio with eventual external collaborators. This dynamic modus operandi improves the answers to the specific demands of each project and blurs some authorial issues, while allowing a permanent change and reinvention of the group, followed by a continuous improvement of its conceptual basis.

The work of each of the five members of the studio takes Brazilian modern architecture as a departure point in different points of view, restating and rethinking some of its concepts to reach adequate and innovative answers to local problems in the realm of architecture and its relation to the city.

Developed in parallel to a docent practice, the work deals with a great range of scales and programmes, from individual houses to public buildings and urban design, always committed to rethink programmatic and construction issues beyond its ordinary sense. (Photographer: João Marcos Ros)

MALAKA ACKAOUI

Malaka Ackaoui is a principal of WAA Inc. She graduated as a bachelor in Landscape Architecture in 1978 (University of Montreal) and became a Master of Urban Planning in 1986 (McGill University). As a Landscape Architect and Urban Planner, Malaka has led several complex and challenging projects. She works both nationally and internationally, namely in China, where her company, WAA has a branch office in Shanghai. She is presently directing the planning and landscape plans for a 350 ha Agrifood Technopark in the province of Yunnan in China.

Malaka's work combines planning, urban design and landscape architecture in a holistic manner and integrates sustainability as a fundamental component of planning and landscape design projects. She directed the design of several institutional projects on the campuses of major Montreal Universities, namely McGill and University of Montreal. Her most recent completed project, James Square at McGill University, combines environmental, artistic and social components along with complex technical challenges.

PEPE GASCÓN

Pepe Gascón was born in Barcelona, November 1973, qualified as an Architect at ESARQ (Universitat Internacional de Catalunya) in 2002, qualified as an Architectural Technologist at the EUPB (Universitat Politècnica de Catalunya) in 1997.

Pending Engineering end-of-course project in Industrial Organisation at the EPSEB (Universitat Politècnica de Catalunya).

In 2002 and 2003 he worked as an assistant to Professor Ignasi Pérez and Doctor in Architecture Anna Vincenza Nufrio at ESARQ (Universitat Internacional de Catalunya) in the subjects of Ecology and Cooperation respectively.

In 2003 he established Pepe Gascón Arquitectura, a multidisciplinary consulting architecture that has a young and innovative team, which develops projects of architecture, urbanism and interior design.

His work is oriented to both the public and private, and includes projects of varying scale and technical complexity.

His work has been published and exhibited in numerous national and international publications and exhibitions.

SCAPE LANDSCHAFTSARCHITEKTEN GMBH

For the scape office the term "landscape" comprises not only the area of unspoiled nature outside the cities, but all forms of landscape like cityscapes, seascapes and industrial landscapes.

These landscapes are not naturally available, but emerge when a partition of the environment is realized aesthetically. As designers they want to induce, support and direct the processes, which cause development of landscape. The term "scape" stands for designed landscape in the broadest sense: land becomes landscape, city becomes urbanscape, depending on the object, materiality and dimension. If a project shall give impulses for the image of a location, it cannot just satisfy basic requirements. They aim at developing projects for their customers, which offer an outstanding design and allow a diversity of adoptions. Sustainable high-tech solutions result from the use of innovative and ecological methods of construction.

CHYUTIN ARCHITECTS ltd.

Chyutin Architects was founded by Bracha Chyutin and Michael Chyutin in 1978. The firm specializes in architectural design of public buildings, residential buildings and interiors, and in urban planning. The firm has been responsible for the design of the master plans for several academic campuses and buildings, as well as for many culture and education complexes. The firm's designs have been shown at numerous national and international architectural exhibitions, and have been written about in the daily press and in professional journals throughout the world. Bracha Chyutin and Michael Chyutin still engage in academic activity in most of Israel's universities. They have published a large number of papers and studies about architecture and urban planning, and have co-authored several books.

PWP LANDSCAPE ARCHITECTURE

PWP Landscape Architecture has six partners and 35 employees. Based in Berkeley, California, the partners, associates, and staff bring a range of interests and expertise to their shared ambition in creating legible, beautiful, and enduring places. They combine knowledge of history and tradition with fluency in contemporary landscape debate. They bring their mastery of construction to bear on the latest technologies and innovations. They draw on their extensive experience with project delivery in some of the world's most politically and physically complex environments to address the challenges of each project. Through constant interchange with clients and architects and the effective management of numerous and diverse consultants, PWP produces landscape architecture of the highest caliber.

BURO LUBBERS

Buro Lubbers is a design studio for landscape architecture, urban development and the design of public space. Since 1993 the studio has been designing projects at all scale levels, for both private and public commissioners. The studio's projects are characterised by a multidisciplinary approach, innovative expertise, sharp analysis, a sturdy and poetic expression and a critical eye for detail. An enthusiastic team takes care of a total product, from research to preservation.

LABICS

Based in Rome, Labics is an architectural and urban planning practice led by Maria Claudia Clemente and Francesco Isidori. Founded in 2002, Labics work across projects of widely varying scales, from small interior design schemes to large urban masterplans. The practice's international portfolio of built projects is backed up by a strong research base and a philosophy which balance pragmatism with lyricism.

ERNEST FERRÉ RICART & MAMEN DOMINGO DOMINGO

Ernest Ferré Ricart and Mamen Domingo Domingo have been a professional team since 1992.

Ernest Ferré is a professor in urbanism and studio projects at the "Escola d'Arquitectura de Reus" (School of Architecture of Reus). He is also a professor in the master studies "The Culture of the Metropolis_UPC". He has been a member of the commission for Architecture Quality of the Barcelona city council.

Mamen Domingo is a professor in studio projects at ETSAB. Besides, she had been a guest professor at the University of Architecture ETH Zurich for two years. She is a member of the commission for Architecture of the PEMB Metropolitan Strategic Plan of Barcelona.

www.domingoferre.com

ALLESWIRDGUT

Founder: Andreas Marth, Christian Waldner, Friedrich Passler, Herwig Spiegl

Since 1997, AllesWirdGut has been working on projects of different scales – from urban strategies to interior design. The pragmatic approach searches for the potential of the given context. By looking at so-called problems as chances, new unexpected possibilities come into existence. The goal beyond the given task is to find additional qualities in order to realise them. The four architects of AllesWirdGut met at the Technical University of Vienna (TU) where first common projects were created. What distinguishes the team is the input of diversified characters as well as the co-operation without hierarchies and without specialisations.

The actual formula for success of the young architects is team spirit. The building task is analised thoroughly, nothing is forbidden on principle, and everything is possible. The logic answer at first sight is not always the best. As a result surprising methods of resolutions occur for both the builders and the architects. Those methods are reviewed and perfected mutualy.

AllesWirdGut puts its emphasis on contents and synergies – on resources that complement one another and therefore serve economic functions. Architecture is not an end in itself, the stress is laid on the benefit. Good architecture should not be more expensive – but simply be more efficient!

METROBOX ARCHITEKTEN

Metrobox architekten was founded in 2005 in Cologne, Germany. The office is following a holistic design intent exploring the boundaries between architecture, urban and landscape design as well as interior furniture design. The design process goes along with intensive research work as well as strong integration of involved groups of users. These strategies lead to unique results fitting perfectly the customers' needs and – in this case – help creating livable public spaces.

LEVIN MONSIGNY LANDSCAPE ARCHITECTS

Levin Monsigny Landscape Architects is a French-German landscape design firm based in Berlin. They offer a full range of landscaping and urban design and have carried out commissions for public plazas and parks, mixed-use developments, residential estates, private and historical gardens. Their highly-experienced international team is united by its passion for discovering new places, for contributing to a more livable environment and by the challenge of finding individual solutions to match people's needs. The unusual results have the stamp of an unmistakable identity of place. The basis of every design is an exact analysis of the site within its surroundings and the knowledge of its history and traditions, as well as an optimal organisation of the nature of the task. They are involved in each project from the initial sketches to the concrete on-site execution. They are thus able to guarantee a design of the highest possible quality.

RUUD-JAN KOKKE

The designs of Ruud-Jan Kokke (1965 Velp, The Netherlands) are characterized by a personal style and design with special attention to detail and functionality. They have a high technical and visual quality with a sober and inventive, playful character.

Kokke started his own studio in 1986 and puts his energy into finding solutions that enhance the ease of use. Kokke focuses on Furniture design, Architectural and Interior design projects, as well as projects in Public areas.

His work is published in the prominent Design Yearbook 1998. Some of his designs are exhibited in the Museum of Modern Art in New York, the "Stedelijk Museum" in Amsterdam and the "Museum für Angewandte Kunst" in Cologne.

www.kokke.com

SARA TAVARES COSTA AND PABLO DÍAZ-FIERROS

Sara Tavares Costa graduated from University of Porto (FAUP 1997) and Pablo Díaz-Fierros graduated from University of Seville (ETSA Seville 2002).

After collaborating for years in Guillermo Vázquez Consuegra studio, they established COSTA FIERROS ARQUITECTOS in 2004, located in Seville.

In 2009 they were selected by "Fundación de Arquitectura Contemporánea" (Architectural Contemporary Foundation) as one of the eight young andalusian architects studios contributing with new experiences in housing and contemporary city design.

They have participated in different architectural competitions in many areas. Their projects have been published in national and international magazines.

Individually Sara Tavares Costa works in architectural projects with other architects, engineers, industrial designers, biologists, etc. Pablo Díaz-Fierros is also a professional architectural photographer (www.pablodiazfierros.com).

ASPECT STUDIOS

ASPECT Studios is a creative landscape architecture design studio committed to providing positive outcomes for people and places through contemporary, innovative and sustainable design. The Chinese studio is based in Shanghai.

Sustainability is a core value of the Studio. Sustainability is a complex interplay of cultural, environmental, social and economic elements and they seek to balance these elements through a holistic design approach. This is demonstrated through numerous awards and the quality of their built and strategic work.

The Directors of ASPECT Studios are respected professionals in Australia, having lectured at universities throughout the world.

Solid working experience in China allows them to interpret local culture in a contemporary manner through a sustainable design approach. The knowledge of local materials and planting, their passion for new technologies also contribute to the successful completion of their projects.

GUSTAFSON PORTER

Gustafson Porter's practice has developed an international reputation for creating innovative and contemporary landscape design. Their work is known for its very sensual approach to landscape design. They aim to distil the soul of the place and manifest this in the physical design. Their core strength lies in their design skills and the ability to deliver high-quality design on high-profile projects. Completed sites include Westergasfabriek Park in Amsterdam, the interior landscape of the Great Glass House at the National Botanic Garden of Wales, two courtyards for the Government Offices, the Swiss Cottage Open Space and the Diana, Princess of Wales Memorial Fountain. Old Market Square in Nottingham is the UK's second largest public square after Trafalgar Square and has received numerous awards for design quality within the public realm. These projects established Gustafson Porter's solid reputation for designing and executing high quality contemporary landscapes within often complex historical contexts.

LANDSCAPE PROJECTS

Landscape Projects is a landscape and urban design practice, recognised as a leader in its field. It was established by Neil Swanson CMLI, whom with his associate director Rachel Devine CMLI manage a workload of prestigious projects across the UK.

At the heart of the practice's work is a concern to make the contemporary city a better place to live, work and relax. In its completed projects and current work the practice explores new ways to connect people and landscape through design.

Landscape Projects has established a reputation for high quality public realm design and has won competitions and awards for its work, including Civic Trust Awards, CABE and Design Council recognition of good practice in landscape design, and international design competitions in the Europe and the UK.

This reputation has translated into a wide-ranging portfolio of prestigious projects and clients. The practice has gained its workload entirely by word of mouth, and well over half of Landscape Projects commissions come repeat work from satisfied clients, a major testimonial to the reputation of the practice.

BENEDETTA TAGLIABUE

Benedetta Tagliabue was born in Milan and graduated from the University of Venice in 1989. In 1991 she joined Enric Miralles' studio where she eventually became a partner.

In 2012 she taught at Graduate School of Design Harvard University. In 2010 she taught at Columbia University in New York and she was a workshop professor at TEC de Monterrey (Mexico) the School of Architecture in Venice (IUAV) and at École Speciale d'Architeture (ESA, Paris). Besides being a lecturer she attended nearly once a month in architectural forums all over the world, such as the RIBA, the Architectural Association and the Bartlett School of Architecture in London, the Berlage Institute in Amsterdam and at Columbia University in New York, among others.

Recently she received the RIBA's International Fellowships for her particular contribution as a non-UK architect she has made to architecture (2009) and an Honorary Doctor of Arts degree from Napier University (2004). She was invited to participate as a jury in international architectural competitions and prestigious awards including the RIBA Stirling Prize 2009.

VIBEKE RØNNOW

Year of birth: 1950

Nationality: Danmark

Profession: Landscape architect MDL, Architect MAA

Position: Head of department, C.F. Møller Landscape

Professional career: C. F. Møller Architects since April 4th 2008, on the takeover of Vibeke Rønnow Landscape Architects

1987–2008: Own practice, Vibeke Rønnow Landscape Architects Aps

Awards: The Lighting Prize 2011, Aalborg Waterfront. Aarhus Municipality Architectural Prize 2011, Sundby Wharf. Aarhus Municipality Architectural Prize 2011, Aalborg Waterfront.

BURO SANT EN CO

Buro Sant en Co was founded in 1990 by Monique de Vette and Edwin Santhagens. The office provides a full range of landscape architecture and urban design services, conceived and executed at the highest artistic level.

Employing a team of skilled designers, each with a different specialization, the company is able to compete and take part in projects at every level, from masterplan to detailed design. The result is a large variety of realized projects such as parks, campuses, squares, shopping areas, boulevards, gardens and estates.

Designing is the search for distinctive and sensitive solutions to individual problems. The style of design is important. They aim to make innovative plans that have a certain beauty and artistic expression. Yet functionality and maintenance are just as important. They spent great efforts on detailing and material choice. Functionality, simplicity and sustainability are keywords for their design approach.

BATLLE I ROIG ARQUITECTES

Batlle i Roig Arquitectes is an architectural practice founded in 1981 by Enric Batlle and Joan Roig, formed by a multidisciplinary team able to reach all areas of architecture, landscape, environment, sustainability, engineering and technical architecture.

The goal of the firm is the architecture and the client, to whom they offer their service in the broadest sense, having as general guidelines when making their projects the social and cultural commitment, the technological innovation, creation and respect for the environment.

The practice consists of 40 professionals of all areas of architecture, landscape, engineering and technical architecture. They also have a number of external collaborators with an important career and extensive experience in each area, a working system that allows the whole to respond quickly and reach a capacity of commitment to their clients and their projects.

GILLESPIES

Gillespies is a leading international masterplanning, landscape architecture and urban design firm. Their international reputation has been built upon creative design, and a track record for high quality projects.

Gillespies tackles some of the largest and most complex commissions available to the profession.

Every new design creates a legacy for the future, and in a globalised world, places exist and compete on many levels. Gillespies designs places with individual identity, which deliver renewed economic prosperity, social and cultural enrichment and enhanced quality of life. Sustainable design is also central to their work.

Implemented projects include designs for residential, mixed-use, offices and commercial, leisure, tourism and sports developments, education and healthcare, transport and infrastructure projects, and high-profile public realm schemes.

Finished projects are located in every major UK city, Western and Eastern Europe, Asia, Middle East and North Africa, the Americas and Australia.

Designs by the Gillespies team have been recognised by numerous industry awards. The Gillespies team has the resources for all scales of projects, in various locations.

www.gillespies.co.uk

ATELIER DE PAYSAGES BRUEL-DELMAR

Funder: Christophe Delmar and Anne-Sylvie Bruel

Atelier de Paysages Bruel-Delmar are landscape architects but also town planners. Their studio is in Paris but they act trough all French territory. On the side, Christophe and Anne-Sylvie are a lso teaching in different architecture or landscape schools in order to share their experience.

After their first solutions for extreme sites, the Atelier de Paysages Bruel-Delmar undertook long-term work on urban transformation projets where the revealed and rediscovered territory weaves the link with new developments. They actually work for major French towns on sustainable district and take great care of water remediation.

Bruel & Delmar has won several prestigious prizes in France in the latest years such as the first Price for Urban Development in 2010 and the Great Prize for Water features in townscape in 2011.

后记

本书的编写离不开各位设计师和摄影师的帮助，正是有了他们专业而负责的工作态度，才有了本书的顺利出版。参与本书的编写人员有：

Bracha Chyutin, Michael Chyutin, Ethel Rosenhek, Joseph Perez, Jacques Dahan, Sharon yeari, Maria Claudia Clemente, Francesco Isidori, Labics, Luigi Filetici, Gillespies, Giles Rocholl, SARA TAVARES COSTA, PABLO DÍAZ-FIERROS, PABLO DÍAZ – FIERROS, Javier Fresneda Puerto, Javier Sanjuán Calle (MTM architects), Javier Peña Galiano (Xpiral architect), Miralles Tagliabue EMBT, ALEX GAULTIER, ASPECT Studios, FJMT, Florian Groehn, John Marmaras, Fernando G. Pino, Manuel G. de Paredes, Paredes & Pino Architects, Paredes Pino, Mamen Domingo, Ernest Ferré, architects, Jordi Bernadó, Cristobal Bryjowski, ADEPT, Kaare Viemose, AllesWirdGut, Bjørbekk & Lindheim AS, György Szloszjár, Borbála Salgó, Eszter Dandé, Garten Studio Ltd. Wa, Attila Polgár, Garten Studio Ltd., Malaka Ackaoui, Williams, Asselin, Ackaoui & Associates Inc., Vincent Asselin and Groupe TREMCA Inc., Ronald Bron, Frits van Loon, Elizabeth Keller, Petrouschka Thumann, Marike Oudijk, HOSPER, Pieter Kers, Edwin Santhagens, Stephen Tas, Buro Sant en Co, Bjørbekk & Lindheim, Div`A arkitekter, Hendrik Bruns, Jens Bartsch, Jan Sauerborn, Metrobox Architekten, Peter Seynaeve, Tompy Hoedelmans, Koen Moelants, Luc Wallays, Evi Lefevere, OMGEVING, OLM, Levin Monsigny Landscape Architects, Claas Dreppenstedt, Berlin and Arial view - DOM publishers, Berlin, Claude Cormier + Associés Inc. / Consortium Cardinal Hardy, Consortium Cardinal Hardy, Claude Cormier + Associés inc, Denis Tremblay, Marc Cramer, Atelier de Paysages BRUEL-DELMAR, Max Lerouge & Atelier de paysage Bruel – Delmar, Alexandre Brasil, André Luiz Prado, Bruno Santa Cecília, Carlos Alberto Maciel, Eduardo Eckenfels, Pepe Gascón, Landscape Projects, Shaw & Shaw, Landscape Projects, Matthias Funk, Hiltrud M. Lintel, Rainer Sachse, scape Landschaftsarchitekten GmbH, Hiltrud M. Lintel, Vibeke Rønnow, C. F. Møller Architects, Hanneke Kijne, Petrouschka Thumann, Han Konings, Remco Rolvink, Ronald Bron, Hilke Floris, Bureau B+B stedebouw en landschapsarchitectuur, Renee Klein, Frederica Rijkenberg, Peter Walker, FASLA, PWP Landscape Architecture, Enric Batlle, Joan Roig, Batlle i Roig Arquitectes, Ruud-Jan Kokke, Gabriele Merolli, Buro Lubbers, Gustafson Porter, Gustafson Porter, Grant Smith, Chris Mansfield and Redshift Photography,

本书译者为大连工业大学外国语学院的何心

ACKNOWLEDGEMENTS

We would like to thank everyone involved in the production of this book, especially all the artists, designers, architects and photographers for their kind permission to publish their works. We are also very grateful to many other people whose names do not appear on the credits but who provided assistance and support. We highly appreciate the contribution of images, ideas, and concepts and thank them for allowing their creativity to be shared with readers around the world.